Frontiers in Probability and the Statistical Sciences

The "Frontiers" is a series of books (edited volumes and monographs) in probability and statistics designed to capture exciting trends in current research as they develop. Some emphasis will be given on novel methodologies that may have interdisciplinary applications in scientific fields such as biology, ecology, economics, environmental sciences, finance, genetics, material sciences, medicine, omics studies and public health.

Nicolas Marie

From Nonparametric Regression to Statistical Inference for Non-Ergodic Diffusion Processes

Springer

Nicolas Marie
Modal'X Department
Paris Nanterre University
Nanterre, France

ISSN 2624-9987 ISSN 2624-9995 (electronic)
Frontiers in Probability and the Statistical Sciences
ISBN 978-3-031-95637-9 ISBN 978-3-031-95638-6 (eBook)
https://doi.org/10.1007/978-3-031-95638-6

Mathematics Subject Classification: 62G05, 60J60, 62M05, 62M09, 60H10

This Springer imprint is published by the registered company Springer Nature Switzerland AG
The registered company address is: Gewerbestrasse 11, 6330 Cham, Switzerland

To my dear Nathalie

Preface

Stochastic Differential Equations (SDEs) are very popular in applications, and statistical inference for these models has been an intensively investigated research field for many years. On estimators computed from one long-time observation of the ergodic stationary solution of an SDE, there are already famous books, such as Kutoyants [1] (for diffusions) and Kubilius et al. [2] (for fractional diffusions). However, to my knowledge, there is no book on copies-based statistical inference for such models, which is related to functional data analysis. So, following a presentation on this topic to my colleagues at the Modal'X Department of Paris Nanterre University, I decided to write the present book on the projection least squares and Nadaraya-Watson estimators of the drift function computed from multiple short-time observations of an SDE solution. The main purpose of this book is to show to the reader how to extend proof techniques from nonparametric regression to the copies-based estimation of the drift function for various usual models: SDEs driven by Brownian motion, a Lévy process or fractional Brownian motion. The story doesn't stop there, because I'm already working on a second book on copies-based nonparametric estimators of the drift function for more complicated models, and of the transition density function for diffusion processes. Anyway, I hope this first book will help the reader to become familiar with copies-based statistical inference for SDE models.

Finally, I would like to thank my main co-author Fabienne Comte for all her valuable comments on the manuscript.

Paris, France
2025

Nicolas Marie

References

1. Kutoyants, Y. (2004). *Statistical Inference for Ergodic Diffusion Processes*. Springer, London.
2. Kubilius, K., Mishura, Y. and Ralchenko, K. (2017). *Parameter Estimation in Fractional Diffusion Models*. Bocconi & Springer Series.

Competing Interests The author has no competing interests to declare that are relevant to the content of this manuscript.

Contents

Chapter 1
Introduction

A Stochastic Differential Equation (SDE) is a differential equation perturbed by a random component, defined thanks to an integral with respect to a stochastic process such as Brownian motion, a jump process and fractional Brownian motion. When the driving signal is Brownian motion, the stochastic integral is taken in the sense of Itô, and the solution of the SDE is a diffusion process (see Revuz and Yor [1]). These are widely used to model random dynamical systems in applications: the prices of a risky asset in mathematical finance (see Lamberton and Lapeyre [2]), the elimination process of a drug in pharmacokinetics (see Donnet and Samson [3]), the membrane potential of a neuron (see Bachar et al. [4]), etc. So, in order to fit stochastic differential equations driven by Brownian motion on data in all these applications, one uses statistical inference for diffusion processes, which has been an intensively investigated research field for many years. Precisely, from continuous-time observations, the drift function b_0 of the SDE needs to be estimated. Until the end of the 1990s, all the estimators of b_0 were computed from one long-time observation of the ergodic stationary solution of the SDE. More recently, new kinds of estimators of the drift function have been investigated: those computed from multiple short-time observations of the diffusion process. In both approaches, the estimators of b_0 studied in the literature are derived from simpler estimators involved in the (non)parametric regression framework: the maximum likelihood estimator, the (projection) least squares estimator, kernel-based methods, etc. In some situations, Brownian motion is not the appropriate driving signal, for instance to model the prices process of a risky asset when jumps may occur, or to model its volatility process whose memory and path regularity may differ from those of a diffusion process. Such random dynamical systems are modeled by stochastic differential equations driven by a Lévy process (see Applebaum [5]) or by fractional Brownian motion (see Friz and Victoir [6]). Here again, long-time behavior-based estimators and copies-based estimators of b_0 have been investigated in the literature, but additional difficulties need to be managed, especially when the driving signal is fractional Brownian motion, which is not a semi-martingale when its Hurst parameter differs from $1/2$. This book deals with two

N. Marie, *From Nonparametric Regression to Statistical Inference for Non-Ergodic Diffusion Processes*, Frontiers in Probability and the Statistical Sciences,
https://doi.org/10.1007/978-3-031-95638-6_1

types of copies-based estimators of the drift function: the projection least squares estimator for both diffusion processes and stochastic differential equations driven by a Lévy process or by the fractional Brownian motion, and the Nadaraya-Watson kernel-based estimator for diffusion processes.

Let $X = (X_t)_{t\in\mathbb{R}_+}$ be the diffusion process defined by

$$X_t = x_0 + \int_0^t b_0(X_s)ds + \int_0^t \sigma(X_s)dW_s;\ t \in \mathbb{R}_+, \tag{1.1}$$

where $x_0 \in \mathbb{R}$, $W = (W_t)_{t\in\mathbb{R}_+}$ is a Brownian motion, and $b_0, \sigma \in C^1(\mathbb{R})$ have bounded derivatives in order to ensure the existence and the uniqueness of the (strong) solution of Eq. (1.1).

As mentioned above, the oldest kind of (nonparametric) estimators of the drift function b_0 is based on the long-time behavior of the solution of Eq. (1.1). Precisely, they are computed from one observation of the ergodic stationary solution of Eq. (1.1), which exists and is unique under the following conditions on b_0 and σ:

$$\exists d \geqslant 0, \exists c > 0, \exists R \geqslant 0 : \forall x \in \mathbb{R}, |x| \geqslant R \Longrightarrow xb_0(x) \leqslant -c|x|^{d+1} \tag{1.2}$$

and

$$\exists \alpha, A > 0 : \forall x \in \mathbb{R}, \alpha \leqslant |\sigma(x)| \leqslant \alpha + A. \tag{1.3}$$

The reader may refer to Hoffmann [7] on a discrete-time wavelet-based estimator, to Kutoyants [8], Chap. 4 or Dalalyan [9] on continuous-time kernel-based estimators, to Comte et al. [10] on a discrete-time projection least squares estimator, etc. Note that b_0 satisfies (1.2) when $b_0(0) = 0$ and

$$\exists c > 0 : \forall x \in \mathbb{R}, b_0'(x) \leqslant -c. \tag{1.4}$$

Indeed, for $x \geqslant 0$ (resp. $x < 0$), by the mean value theorem and since $b_0(0) = 0$, there exists $u \in (0, x)$ (resp. $u \in (x, 0)$) such that $b_0(x) = b_0'(u)x$, leading to

$$xb_0(x) = x^2 b_0'(u) \leqslant -cx^2.$$

So, for instance, it is clear that the drift and volatility functions of the so-called Langevin equation $dX_t = -\theta X_t dt + dW_t$ with $\theta > 0$ satisfy (1.2)–(1.3). But what about the estimation of θ when it belongs to $(-\infty, 0)$? To answer this question, and many others, a new kind of estimators of b_0 has been investigated since several years: those computed from N copies $X^1, \dots, X^N$ of X observed on $[0, T]$ with $T > 0$ fixed but $N \to \infty$. The major part of the literature deals with estimators based on independent copies of X (see Comte and Genon-Catalot [11] on a continuous-time projection least squares estimator, Denis et al. [12] on a discrete-time nonparametric ridge estimator, Marie and Rosier [13] on both continuous-time and discrete-time versions of a Nadaraya-Watson estimator, etc.). However, some recent papers are also devoted to estimators based on dependent copies. For instance, Della Maestra and Hoffmann

[14] deals with a continuous-time Nadaraya-Watson estimator in interacting particle systems, and Comte and Marie [15] deals with a projection least squares estimator computed from multiple copies of X driven by correlated Brownian motions.

With copies-based estimators, the drift function b_0 doesn't need to satisfy (1.2), and for instance, θ may be negative in the previous Langevin equation. Moreover, these estimators are well-adapted to some situations difficult to manage with long-time behavior-based estimators:

- Assume that X models the elimination process of a drug administered to one people, and assume that in a clinical-trial involving N patients, X^i models the elimination process of the same drug for the i-th patient. Then, $X^1, \dots, X^N$ are independent copies of X and the second kind of estimators of b_0 is tailor-made in such situation.
- Consider N interacting risky assets of same kind such that the prices process X^i of the i-th asset is modeled by $dX_t^i = b_0(X_t^i)dt + \sigma(X_t^i)dW_t^i$, where $W^1, \dots, W^N$ are correlated Brownian motions (see Duellmann et al. [16]). Although $X^1, \dots, X^N$ are not independent, the second kind of estimators of b_0 remains appropriate in this situation. In order to provide a very simple example of copies-based parametric estimator of b_0 in this financial setting, assume that $dX_t^i = \theta_0 X_t^i dt + X_t^i dW_t^i$, and that

$$d\langle W^i, W^k\rangle_t = R_{i,k}dt;\ \forall i, k \in \{1, \dots, N\},$$

where R is a $N \times N$ correlation matrix. For every $t \in [0, T]$,

$$X_t^i = x_0 e^{Y_t^i} \quad \text{with} \quad Y_t^i = \left(\theta_0 - \frac{1}{2}\right)t + W_t^i.$$

A natural estimator of θ_0 is given by

$$\widehat{\theta}_N = \frac{1}{2} + \frac{1}{NT}\sum_{i=1}^{N} Y_T^i = \theta_0 + \frac{1}{NT}\sum_{i=1}^{N} W_T^i,$$

and its quadratic risk is easy to compute

$$\begin{aligned}\mathbb{E}((\widehat{\theta}_N - \theta_0)^2) &= \frac{1}{N^2T^2}\left(\sum_{i=1}^{N}\mathbb{E}(|W_T^i|^2) + \sum_{i\neq k}\mathbb{E}(W_T^i W_T^k)\right)\\ &= \frac{1}{NT}\left(1 + \frac{1}{N}\sum_{i\neq k} R_{i,k}\right) \xrightarrow[N\to\infty]{} 0 \quad \text{when} \quad \sum_{i\neq k} R_{i,k} \underset{N\to\infty}{=} \mathrm{o}(N^2).\end{aligned}$$

Finally, from one path of the solution X to Eq. (1.1) observed on $\mathbb{R}_+$, which seems to be a situation only appropriate for long-time behavior-based estimators of b_0, one can construct N independent copies of $X_{|[0,T]}$ when b_0 and σ satisfy (1.2)–(1.3). To that purpose, consider the stopping times $\tau_1, \dots, \tau_N$ recursively defined by $\tau_1 = 0$ and

$$\tau_i = \inf\{t > \tau_{i-1} + T : X_t = x_0\}; i = 2, \dots, N$$

with the convention $\inf(\emptyset) = \infty$. Since b_0 and σ fulfill (1.2)–(1.3), the scale density

$$s(\cdot) = \exp\left(-2\int_0^{\cdot} \frac{b_0(x)}{\sigma(x)^2}dx\right)$$

satisfies

$$\int_{-\infty}^0 s(x)dx = \int_0^{\infty} s(x)dx = \infty,$$

and then X is a recurrent Markov process by Khasminskii [17], Example 3.10. So, for any $i \in \{1, \dots, N\}$, $\mathbb{P}(\tau_i < \infty) = 1$, and one can consider the processes

$$W^i = (W_{\tau_i+t} - W_{\tau_i})_{t\in[0,T]} \quad \text{and} \quad X^i = (X_{\tau_i+t})_{t\in[0,T]}.$$

Since $W^1, \dots, W^N$ are independent Brownian motions by the strong Markov property, and since

$$X^i = \mathcal{I}(x_0, W^i); \forall i \in \{1, \dots, N\},$$

where $\mathcal{I}$ is the solution map for Eq. (1.1); $X^1, \dots, X^N$ are independent copies of $X_{|[0,T]}$.

Our book focuses on copies-based continuous-time nonparametric estimators of b_0. Note that for continuous-time observations, to determine the volatility function σ is not a statistical problem. Indeed,

$$\langle X\rangle_t = \int_0^t \sigma(X_s)^2 ds; \forall t \in [0, T],$$

and since X has continuous paths, there exists a random interval I of $\mathbb{R}$ such that for any $x \in I$, one can find $t(x) \in [0, T]$ satisfying $X_{t(x)} = x$, leading to

$$\sigma(x) = \sigma(X_{t(x)}) = \left(\frac{d}{dt}\langle X\rangle_t\Big|_{t=t(x)}\right)^{\frac{1}{2}}.$$

The estimators of b_0 based on copies of the solution of Eq. (1.1) are derived from simpler estimators involved in nonparametric regression. So, consider the i.i.d. random variables $(\xi_1, Y_1), \dots, (\xi_N, Y_N)$ such that, for every $i \in \{1, \dots, N\}$,

$$Y_i = b_0(\xi_i) + \varepsilon_i, \tag{1.5}$$

where the errors $\varepsilon_1, \dots, \varepsilon_N$ are centered and independent of $\xi_1, \dots, \xi_N$. There are two extensions of the so-called (parametric) least squares estimator to the nonparametric regression framework, and then to the copies-based estimation of the drift function for diffusion processes: the projection least squares estimator and the Nadaraya-Watson estimator.

The projection least squares estimator. Let $(\varphi_1, \dots, \varphi_m)$ be an I-supported orthonormal family of $\mathbb{L}^2(\mathbb{R}, dx)$, where $m \in \{1, \dots, N\}$ and I is an interval of $\mathbb{R}$. Consider also $\mathcal{S}_m = \text{span}\{\varphi_1, \dots, \varphi_m\}$, and assume that the φ_j's are continuous on I. It is natural to estimate b_0 (on I) by the element of $\mathcal{S}_m$ minimizing the objective function $\gamma_N^{\text{l.s.}}$ defined by

$$\gamma_N^{\text{l.s.}}(b) = \frac{1}{N}\sum_{i=1}^{N}(Y_i - b(\xi_i))^2 = \gamma_N^{\mathfrak{r}}(b) + \frac{1}{N}\sum_{i=1}^{N} Y_i^2,$$

where

$$\gamma_N^{\mathfrak{r}}(b) = \frac{1}{N}\sum_{i=1}^{N}(b(\xi_i)^2 - 2b(\xi_i)Y_i) \quad \text{for every} b \in \mathcal{S}_m.$$

The objective function $\gamma_N^{\text{l.s.}}$ is a natural extension of the (parametric) least squares criterion. Note also that to minimize $\gamma_N^{\text{l.s.}}$ or $\gamma_N^{\mathfrak{r}}$ on $\mathcal{S}_m$ is equivalent because $\gamma_N^{\text{l.s.}}(b) - \gamma_N^{\mathfrak{r}}(b)$ doesn't depend on $b \in \mathcal{S}_m$. This remark is crucial because $\gamma_N^{\mathfrak{r}}$ can be easily extended to the estimation of b_0 from copies of the diffusion process X, but not $\gamma_N^{\text{l.s.}}$.

Now, assume that $m \in \{1, \dots, N_T\}$ with $N_T = [NT] + 1$, and let γ_N be the objective function defined by

$$\gamma_N(b) = \frac{1}{NT}\sum_{i=1}^{N}\left(\int_0^T b(X_s^i)^2 ds - 2\int_0^T b(X_s^i)dX_s^i\right) ; \forall b \in \mathcal{S}_m.$$

Let us show that γ_N has a unique minimizer in $\mathcal{S}_m$. For $b = \sum_{j=1}^{m} \theta_j\varphi_j$ with $\theta_1, \dots, \theta_m \in \mathbb{R}$,

$$\begin{aligned}\nabla\gamma_N(b) &= \left(\frac{2}{NT}\sum_{i=1}^{N}\left[\int_0^T \varphi_j(X_s^i)b(X_s^i)ds - \int_0^T \varphi_j(X_s^i)dX_s^i\right]\right)_{j\in\{1,\dots,m\}} \\ &= 2(\widehat{\Psi}_m\theta - \widehat{Z}_m),\end{aligned}$$

where $\theta = (\theta_1, \dots, \theta_m)$, $\widehat{\Psi}_m = (\langle\varphi_j, \varphi_\ell\rangle_N)_{j,\ell}$, $\langle\cdot, \cdot\rangle_N$ is the empirical inner product defined by

$$\langle\varphi, \psi\rangle_N := \frac{1}{NT}\sum_{i=1}^{N}\int_0^T \varphi(X_s^i)\psi(X_s^i)ds$$

and

$$\widehat{Z}_m = \left(\frac{1}{NT}\sum_{i=1}^{N}\int_0^T \varphi_j(X_s^i)dX_s^i\right)_{j\in\{1,\dots,m\}}.$$

The symmetric matrix $\widehat{\Psi}_m$ is positive semidefinite because, for every $x \in \mathbb{R}^m$,

$$x^*\widehat{\Psi}_m x = \frac{1}{NT}\sum_{i=1}^{N}\int_0^T \left(\sum_{j=1}^{m} x_j\varphi_j(X_s^i)\right)^2 ds \geqslant 0.$$

If in addition $\widehat{\Psi}_m$ is invertible, it is positive definite, and then

$$\widehat{b}_m = \sum_{j=1}^{m}\widehat{\theta}_j\varphi_j \quad \text{with} \quad \widehat{\theta} = \widehat{\Psi}_m^{-1}\widehat{Z}_m \tag{1.6}$$

is the only minimizer of γ_N in $\mathcal{S}_m$, called the projection least squares estimator of b_0.

On the projection least squares estimator, see Comte [18], Chap. 4 in the nonparametric regression framework and, as mentioned above, see Comte and Genon-Catalot [11] (resp. Comte and Marie [15]) in the independent (resp. dependent) copies of diffusion processes framework.

The Nadaraya-Watson estimator. Let K be a kernel, that is an integrable and symmetric function such that

$$\int_{-\infty}^{\infty} K(x)dx = 1,$$

and consider $K_h(\cdot) = K(\cdot/h)/h$ with $h > 0$. In order to extend in a second way the (parametric) least squares criterion to the nonparametric regression framework, let us consider the objective function $G_N(x,\cdot)$ ($x \in \mathbb{R}$) such that

$$G_N(x,\beta) = \frac{1}{N}\sum_{i=1}^{N}(Y_i - \beta)^2 K_h(\xi_i - x);\ \forall \beta \in \mathbb{R}.$$

Clearly,

$$\partial_\beta G_N(x,\beta) = -\frac{2}{N}\sum_{i=1}^{N}(Y_i - \beta)K_h(\xi_i - x),$$

and then the minimizer of $G_N(x,\cdot)$ is

$$\widehat{b}_h^{\mathfrak{X}}(x) = \frac{\sum_{i=1}^{N} K_h(\xi_i - x)Y_i}{\sum_{i=1}^{N} K_h(\xi_i - x)}.$$

The random function $\widehat{b}_h^{\mathfrak{X}} : x \mapsto \widehat{b}_h^{\mathfrak{X}}(x)$ is the Nadaraya-Watson estimator, of bandwidth h, of the regression function b_0 in Model (1.5).

When σ satisfies (1.3), for every $t \in (0, T]$, the probability distribution of X_t has a density f_t with respect to Lebesgue's measure such that $t \mapsto f_t(x)$ ($x \in \mathbb{R}$) belongs to $\mathbb{L}^1([0, T], dt)$ (see Menozzi et al. [19], Theorem 1.2). This legitimates to consider the density function f defined by

$$f(x) = \frac{1}{T - t_0} \int_{t_0}^{T} f_s(x)ds; \forall x \in \mathbb{R},$$

where $t_0 \in [0, T)$. Let us now introduce a Nadaraya-Watson (type) estimator of b_0 computed from copies $X^1, \dots, X^N$ of the diffusion process X:

$$\widehat{b}_h(x) = \frac{\widehat{bf}_h(x)}{\widehat{f}_h(x)}; x \in \mathbb{R}, \tag{1.7}$$

where

$$\widehat{f}_h(x) = \frac{1}{N(T - t_0)} \sum_{i=1}^{N} \int_{t_0}^{T} K_h(X_s^i - x)ds \quad \text{is an estimator of } f(x)$$

and

$$\widehat{bf}_h(x) = \frac{1}{N(T - t_0)} \sum_{i=1}^{N} \int_{t_0}^{T} K_h(X_s^i - x)dX_s^i \quad \text{is an estimator of } b_0(x)f(x).$$

By calculating the expectation of the denominator and that of the numerator of Nadaraya-Watson's estimator, one can roughly show that it seems to be an appropriate estimator of b_0. Indeed, for every $x \in \mathbb{R}$,

$$\begin{aligned}\mathbb{E}(\widehat{f}_h(x)) &= \frac{1}{T - t_0} \int_{t_0}^{T} \mathbb{E}(K_h(X_s - x))ds \\ &= \int_{-\infty}^{\infty} K_h(z - x)f(z)dz = (K_h * f)(x) \xrightarrow[h \to 0]{} f(x)\end{aligned}$$

and, since Itô's integral with respect to the Brownian motion is centered,

$$\begin{aligned}\mathbb{E}(\widehat{bf}_h(x)) &= \frac{1}{T-t_0}\int_{t_0}^{T}\mathbb{E}(K_h(X_s-x)b_0(X_s))ds\\ &= \int_{-\infty}^{\infty}K_h(z-x)b_0(z)f(z)dz = (K_h*(b_0f))(x)\xrightarrow[h\to 0]{} b_0(x)f(x).\end{aligned}$$

On the Nadaraya-Watson estimator, see Comte [18], Chap. 4 in the nonparametric regression framework and, as already mentioned, see Marie and Rosier [13] in the independent copies of diffusion processes framework.

As mentioned above, Eq. (1.1) is well-adapted to model many random dynamical systems in applications. However, in some situations, the Brownian motion is not the appropriate driving signal:

- The solution of Eq. (1.1) has continuous paths, which is (for instance) not appropriate to model the prices of a risky asset when jumps resulting from market crashes or gaps may occur (see Cont and Tankov [20]). In order to model random dynamical systems with jumps, an additional component is required in Eq. (1.1):

$$X_t = x_0 + \int_0^t b_0(X_s)ds + \int_0^t \sigma(X_s)dW_s + \int_0^t \gamma(X_s)d\mathfrak{Z}_s;\ t\in\mathbb{R}_+,$$

  where $\mathfrak{Z} = (\mathfrak{Z}_t)_{t\in\mathbb{R}_+}$ is (for instance) a compensated compound Poisson process independent of W, and $\gamma:\mathbb{R}\to\mathbb{R}$ is a bounded Lipschitz continuous function. The reader may refer to Applebaum [5] on the stochastic calculus for jump processes, and to Schmisser [21] (resp. Halconruy and Marie [22]) on a long-time behavior-based (resp. copies-based) projection least squares estimator of the drift function for stochastic differential equations driven by a Lévy process.
- The Brownian motion is a Markov process with α-Hölder continuous paths for every $\alpha\in(0,1/2)$, which is (for instance) not appropriate to model the volatility of a risky asset (see Comte et al. [23] and Gatheral et al. [24]) and questionable in population pharmacokinetics (see Delattre and Lavielle [25] and Marie [26]). A way to control both the memory and the paths regularity of the solution X of Eq. (1.1) is to replace W by the fractional Brownian motion, that is a centered Gaussian process $B = (B_t)_{t\in\mathbb{R}_+}$ such that

$$\mathbb{E}(B_sB_t) = \frac{1}{2}(s^{2\mathrm{H}} + t^{2\mathrm{H}} - |t-s|^{2\mathrm{H}});\ \forall s,t\in\mathbb{R}_+,$$

  where $\mathrm{H}\in(0,1)$. Indeed, the Hurst parameter H controls the memory of B, which exhibits long-range dependence when $\mathrm{H}>1/2$, and the regularity of its paths in the sense that B has α-Hölder continuous paths for every $\alpha\in(0,\mathrm{H})$ (see Fig. 1.1). Unfortunately, B and (then) X are not semi-martingales when $\mathrm{H}\neq 1/2$. So, it is not possible to integrate with respect to X in the sense of Itô, and then to provide simple extensions of estimators as those defined by (1.6) or (1.7) to the fractional diffusions framework. One can consider the pathwise integral (see Friz and Victoir [6]), which has a role to play but not for statistical purposes because there is no sharp enough control of its moments, or the Skorokhod integral (see Decreusefond [27] and Nualart [28]), which is tailor-made to establish risk bounds on estimators but is not directly computable from an observation of X. However, as detailed in Sects. 4.2.2 and 4.2.3, there is a fixed point strategy to compute some copies-based estimators involving the Skorokhod integral. Since to establish theoretical results on (nonparametric) estimators of the drift function for fractional diffusions is a difficult challenge, the volatility function is assumed to be constant in almost all the publications:

$$X_t = x_0 + \int_0^t b_0(X_s)ds + \sigma B_t; t \in \mathbb{R}_+,$$

where $\sigma \in \mathbb{R}^*$. The reader may refer to Saussereau [29] and Comte and Marie [30] (resp. Comte and Marie [31] and Marie [32, 33]) on long-time behavior-based Nadaraya-Watson (resp. copies-based (projection) least squares) estimators of the drift function for fractional diffusions.
Let us say few words about copies-based estimation for fractional diffusions in population pharmacokinetics and in mathematical finance. On the one hand, as mentioned in Delattre and Lavielle [25], the paths of a diffusion process are a little too rough to realistically model the elimination process C of a drug. In order to bypass this difficulty, as suggested in Marie [26], one could model C by a fractional diffusion of Hurst parameter close to 1. Since multiple independent observations are usually available in population pharmacokinetics, this legitimates to investigate the copies-based statistical inference for fractional diffusions. On the other hand, in the spirit of Comte et al. [23], to model the volatility process σ of a risky asset by a fractional diffusion of Hurst parameter larger than $1/2$ allows to take into account the persistence-in-volatility phenomenon. In practice, the independent-copies-based observation scheme is unrealistic for σ. However, for a fractional non-autonomous extension of the Wiggins stochastic volatility model (see Wiggins [34]), Example 4.2 provides a strategy to construct *low correlated* copies of σ from one long-time observation.

Finally, the price to pay for considering functional estimators of b_0 is that some tuning parameters need to be selected from data in practice. For the projection least squares estimator $\widehat{b}_m$, the tuning parameter is the dimension m of $\mathcal{S}_m$. In order to select m in a finite subset $\mathcal{M}_N$ of $\{1, \dots, N_T\}$, one can minimize $m \mapsto \gamma_N(\widehat{b}_m) + \text{pen}(m)$ in $\mathcal{M}_N$, where $\text{pen}(m)$ is of same order as the variance term in the nonadaptive risk bound on the projection least squares estimator of b_0 (see Massart [35] on model selection). For $\widehat{s}_h = \widehat{f}_h$ or $\widehat{bf}_h$, the tuning parameter is the bandwidth h. In order to select h in a finite subset $\mathcal{H}_N$ of $[h_0, 1]$ with $h_0 \in (0, 1)$, one can minimize $h \mapsto \|\widehat{s}_h - \widehat{s}_{h_0}\|^2 + \text{pen}(h)$ in $\mathcal{H}_N$, where $\text{pen}(h)$ is of same order as the variance term in the nonadaptive risk bound on $\widehat{s}_h$. This is the penalized comparison to overfitting (PCO) method introduced by C. Lacour, P. Massart and V. Rivoirard in [36].

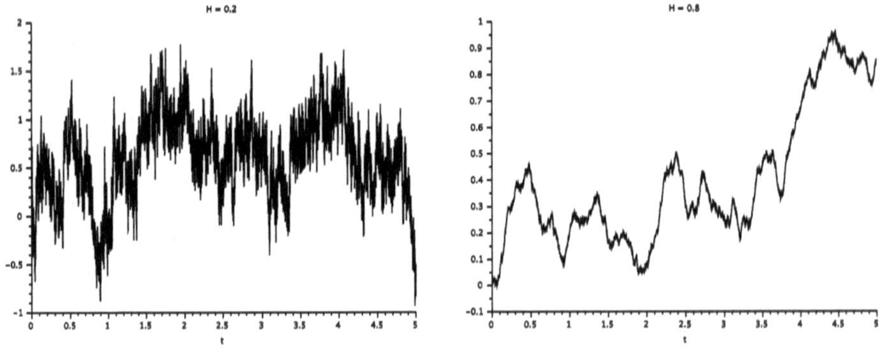

Fig. 1.1 Plots of one path of the fractional Brownian motion for $\mathtt{H} = 0.2$ (left) and $\mathtt{H} = 0.8$ (right)

Chapter 2 is a detailed reminder on nonparametric regression. Chapter 3 deals with risk bounds and model selection for the projection least squares estimator of b_0 defined by (1.6). Both the case of independent and the case of dependent copies of the solution of Eq. (1.1) are investigated. For the reader comfort, brief reminders on Itô's integral and symmetric random matrices are provided. To go further, Chap. 4 deals with the projection least squares estimator of the drift function for stochastic differential equations driven by a jump process first, and then by a fractional Brownian motion. Again, brief reminders on the stochastic integral with respect to a jump diffusion process and on Skorokhod's integral with respect to the fractional Brownian motion are provided. Finally, Chap. 5 deals with risk bounds and bandwidth selection for the Nadaraya-Watson estimator of b_0 defined by (1.7). Precisely, Sect. 5.3 is devoted to the penalized comparison to overfitting bandwidth selection method, in a general framework first, and then applied to our Nadaraya-Watson estimator.

To conclude the introduction chapter, up to our knowledge, this book is the first one dealing with copies-based statistical inference methods for diffusion processes. Such methods are related to the functional data analysis (see Wang et al. [37]) and allow to consider non-ergodic diffusion processes. As mentioned above, the copies-based estimators may be computed from one long-time observation of an ergodic diffusion process, but are also tailor-made for applications where only multiple short-time observations are available. This book focuses on two kinds of copies-based nonparametric estimation methods of the drift function: the projection least squares estimator and a kernel-based estimator. Note that other copies-based statistical methods for SDE, as classification procedures (see Denis et al. [38]), have been investigated but are out of our scope. Another purpose of this book is to consider stochastic differential equations driven by a Lévy process or by the fractional Brownian motion, because such models are widely used in applications, especially in mathematical finance. Finally, having in mind possible applications of the aforementioned estimation methods in Machine Learning, when possible, data-driven procedures are provided.

Notations (general)

- For every $d \in \mathbb{N}^*$, the Euclidean norm on $\mathbb{R}^d$ is denoted by $\|.\|_{2,d}$.
- For every function $\varphi : \mathbb{R} \to \mathbb{R}$, $\|\varphi\|_\infty = \sup_{\mathbb{R}} |\varphi|$.
- The space of the continuous functions from an interval J of $\mathbb{R}$ into $\mathbb{R}$ is denoted by $C^0(J)$.
- The space of the continuously differentiable functions from an interval J of $\mathbb{R}$ into $\mathbb{R}$ is denoted by $C^1(J)$.
- The space of the integrable (with respect to Lebesgue's measure) functions from an interval J of $\mathbb{R}$ into $\mathbb{R}$ is denoted by $\mathbb{L}^1(J, dx)$.
- The space of the square integrable functions from an interval J of $\mathbb{R}$ into $\mathbb{R}$ is denoted by $\mathbb{L}^2(J, dx)$.
- The usual inner product (resp. norm) on $\mathbb{L}^2(\mathbb{R}, dx)$ is denoted by $\langle \cdot, \cdot \rangle$ (resp. $\|.\|$).

References

1. Revuz, D. and Yor, M. (1999). *Continuous Martingales and Brownian Motion*. Springer, Berlin-Heidelberg.
2. Lamberton, D. and Lapeyre, B. (2007). *Introduction to Stochastic Calculus Applied to Finance*. Chapman & Hall/CRC, New York.
3. Donnet, S. and Samson, A. (2013). A Review on Estimation of Stochastic Differential Equations for Pharmacokinetic/Pharmacodynamic Models. *Advanced Drug Delivery Reviews* **65**, 929–939.
4. Bachar, M., Batzel, J. and Ditlevsen, S. (2013). *Stochastic Biomathematical Models with Applications to Neuronal Modeling*. Springer, Berlin-Heidelberg.
5. Applebaum, D. (2009). *Lévy Processes and Stochastic Calculus*. Cambridge University Press, Cambridge.
6. Friz, P. and Victoir, N. (2010). *Multidimensional Stochastic Processes as Rough Paths: Theory and Applications*. Cambridge University Press, Cambridge.
7. Hoffmann, M. (1999). Adaptive Estimation in Diffusion Processes. *Stochastic Processes and their Applications* **79**, 135–163.
8. Kutoyants, Y. (2004). *Statistical Inference for Ergodic Diffusion Processes*. Springer, London.
9. Dalalyan, A. (2005). Sharp Adaptive Estimation of the Trend Coefficient for Ergodic Diffusion. *The Annals of Statistics* **33**, 2507–2528.
10. Comte, F., Genon-Catalot, V. and Rozenholc, Y. (2007). Penalized Nonparametric Mean Square Estimation of the Coefficients of Diffusion Processes. *Bernoulli* **13**(2), 514–543.
11. Comte, F. and Genon-Catalot, V. (2020). Nonparametric Drift Estimation for I.I.D. Paths of Stochastic Differential Equations. *The Annals of Statistics* **48**(6), 3336–3365.
12. Denis, C., Dion, C. and Martinez, M. (2021). A Ridge Estimator of the Drift from Discrete Repeated Observations of the Solutions of a Stochastic Differential Equation. *Bernoulli* **27**, 2675–2713.
13. Marie, N. and Rosier, A. (2023). Nadaraya-Watson Estimator for I.I.D. Paths of Diffusion Processes. *Scandinavian Journal of Statistics* **50**(2), 589–637.
14. Della Maestra, L. and Hoffmann, M. (2022). Nonparametric Estimation for Interacting Particle Systems: McKean-Vlasov Models. *Probability Theory and Related Fields* **182**, 551–613.
15. Comte, F. and Marie, N. (2023). Nonparametric Drift Estimation from Diffusions with Correlated Brownian Motions. *Journal of Multivariate Analysis* **198**, 23 pages.
16. Duellmann, K, Küll, J. and Kunisch, M. (2010). Estimating Asset Correlations from Stock Prices or Default Rates - Which Method is Superior? *Journal of Economic Dynamics & Control* **34**, 2341–2357.

17. Khasminskii, R. (2012). *Stochastic Stability of Differential Equations.* Springer, Berlin-Heidelberg.
18. Comte, F. (2014). *Estimation non-paramétrique.* Spartacus IDH, France.
19. Menozzi, S., Pesce, A. and Zhang, X. (2021). Density and Gradient Estimates for Non Degenerate Brownian SDEs with Unbounded Measurable Drift. *Journal of Differential Equations* **272**, 339–369.
20. Cont, R. and Tankov, P. (2004). *Financial Modeling with Jump Processes.* Chapman & Hall/CRC, Boca Raton.
21. Schmisser, E. (2014). Non-Parametric Adaptive Estimation of the Drift for a Jump Diffusion Process. *Stochastic Processes and their Applications* **124**, 883–914.
22. Halconruy, H. and Marie, N. (2024). On a Projection Least Squares Estimator for Jump Diffusion Processes. *The Annals of the Institute of Statistical Mathematics* **76**(2), 209–234.
23. Comte, F., Coutin, L. and Renault, E. (2012). Affine Fractional Stochastic Volatility Models. *The Annals of Finance* **8**, 337–378.
24. Gatheral, J., Jaisson, T. and Rosenbaum, M. (2018). Volatility is Rough. *Quantitative Finance* **18**(6), 933–949.
25. Delattre, M. and Lavielle, M. (2011). Pharmacokinetics and Stochastic Differential Equations: Model and Methodology. Proceedings of the *Annual Meeting of the Population Approach Group in Europe.*
26. Marie, N. (2014). A Generalized Mean-Reverting Equation and Applications. *ESAIM:PS* **18**, 799–828.
27. Decreusefond, L. (2022). *Selected Topics in Malliavin Calculus.* Bocconi & Springer Series.
28. Nualart, D. (2006). *The Malliavin Calculus and Related Topics.* Springer, Berlin-Heidelberg.
29. Saussereau, B. (2014). Nonparametric Inference for Fractional Diffusion. *Bernoulli* **20**(2), 878–918.
30. Comte, F. and Marie, N. (2019). Nonparametric Estimation in Fractional SDE. *Statistical Inference for Stochastic Processes* **22**(3), 359–382.
31. Comte, F. and Marie, N. (2021). Nonparametric Estimation for I.I.D. Paths of Fractional SDE. *Statistical Inference for Stochastic Processes* **24**(3), 669–705.
32. Marie, N. (2023). Nonparametric Estimation for I.I.D. Paths of a Martingale Driven Model with Application to Non-Autonomous Financial Models. *Finance and Stochastics* **27**(1), 97–126.
33. Marie, N. (2025). On a Computable Skorokhod's Integral Based Estimator of the Drift Parameter in Fractional SDE. *Scandinavian Journal of Statistics* **52**(1), 1–37.
34. Wiggins, J.B. (1987). Option Values under Stochastic Volatility. Theory and Empirical Estimates. *Journal of Financial Economics* **19**(2), 351–372.
35. Massart, P. (2007). *Concentration Inequalities and Model Selection.* Springer, Berlin-Heidelberg.
36. Lacour, C., Massart, P. and Rivoirard, V. (2017). Estimator Selection: a New Method with Applications to Kernel Density Estimation. *Sankhya A* **79**(2), 298–335.
37. Wang, J-L., Chiou, J-M. and Mueller, H-G. (2016). Review of Functional Data Analysis. *Annual Review of Statistics and Its Application* **3**, 257–295.
38. Denis, C., Dion, C. and Martinez, M. (2020). Consistent Procedures For Multiclass Classification Of Discrete Diffusion Paths. *Scandinavian Journal of Statistics* **47**(2), 516–554.

Chapter 2
Nonparametric Regression: A Detailed Reminder

This chapter deals with risk bounds on the Nadaraya-Watson estimator and on the projection least squares estimator of b_0 in the nonparametric regression framework (see Model (1.5)). Although nonparametric regression is not the main topic of our book, instructive sketches of proof with the appropriate references are provided. Section 2.1 deals with kernel-based estimators of the common density $\texttt{f}$ of the ξ_i's and of $b_0\texttt{f}$, and then with the Nadaraya-Watson estimator of b_0. Section 2.2 deals with the projection least squares estimator of b_0. Adaptive versions of these estimators are studied. In particular, beyond the nonparametric regression framework, Sect. 2.1.2 presents the recent penalized comparison to overfitting method introduced by Lacour et al. [1].

2.1 Nonparametric Density Estimation and the Nadaraya-Watson Estimator

Consider

$$\widehat{s}_{\mathrm{K},\ell}(x) = \frac{1}{N}\sum_{i=1}^{N}\mathrm{K}(\xi_i, x)\ell(Y_i);\ x \in \mathbb{R},$$

where $(\xi_1, Y_1), \dots, (\xi_N, Y_N)$ are N independent copies of a random variable (ξ, Y), $\ell : \mathbb{R} \to \mathbb{R}$ is a Borel function and K is a symmetric continuous map from $\mathbb{R}^2$ into $\mathbb{R}$. This is an estimator of the function $s : \mathbb{R} \to \mathbb{R}$ defined by

$$s(x) = \mathbb{E}(\ell(Y)|\xi = x)\texttt{f}(x);\ \forall x \in \mathbb{R},$$

where $\texttt{f}$ is the density with respect to Lebesgue measure of the probability distribution of ξ. In the sequel, $\texttt{f}, s \in \mathbb{L}^2(\mathbb{R}, dx)$ and, for the sake of simplicity, $\widehat{s}_{\mathrm{K},\ell}$ is almost always denoted by $\widehat{s}_{\mathrm{K}}$.

N. Marie, *From Nonparametric Regression to Statistical Inference for Non-Ergodic Diffusion Processes*, Frontiers in Probability and the Statistical Sciences,
https://doi.org/10.1007/978-3-031-95638-6_2

Note that

$$\mathbb{E}(\widehat{s}_{\mathrm{K}}(\cdot)) = \mathbb{E}(\mathrm{K}(\xi,\cdot)\mathbb{E}(\ell(Y)|\xi)) = \int_{-\infty}^{\infty}\mathrm{K}(z,\cdot)s(z)dz,$$

which must be as close as possible to s in $\mathbb{L}^2(\mathbb{R}, dx)$. So, with the notations of Chap. 1, two examples of appropriate symmetric continuous maps from $\mathbb{R}^2$ into $\mathbb{R}$ are given by

$$\mathrm{K}_h : (z,x) \in \mathbb{R}^2 \longmapsto K_h(z-x) \tag{2.1}$$

because

$$\int_{-\infty}^{\infty}\mathrm{K}_h(z,\cdot)s(z)dz = (K_h * s)(\cdot) \xrightarrow[h\to 0]{\mathbb{L}^2} s(\cdot),$$

and by

$$\mathrm{K}_m : (z,x) \in \mathbb{R}^2 \longmapsto \sum_{j=1}^{m}\varphi_j(z)\varphi_j(x) \tag{2.2}$$

because

$$\int_{-\infty}^{\infty}\mathrm{K}_m(z,\cdot)s(z)dz = \sum_{j=1}^{m}\langle s,\varphi_j\rangle\varphi_j(\cdot) \xrightarrow[m\to\infty]{\mathbb{L}^2} s(\cdot).$$

If $\ell = 1$ and K is of type (2.1) (resp. (2.2)), then $\widehat{s}_{\mathrm{K}}$ is the Parzen-Rosenblatt (resp. a projection) estimator of the density function $\mathtt{f}$ (see Comte [2], Chaps. 2 and 3). If $\ell(y) = y$ for every $y \in \mathbb{R}$, and if $(\xi_1, Y_1), \dots, (\xi_N, Y_N)$ satisfy (1.5), then $\widehat{s}_{\mathrm{K}}$ is an estimator of $b_0\mathtt{f}$, and

$$\widehat{b}_{\mathrm{K}} = \frac{\widehat{s}_{\mathrm{K},\ell}}{\widehat{s}_{\mathrm{K},1}} \quad \text{is the Nadaraya-Watson estimator of } b_0.$$

In Sect. 2.1.1, a risk bound on $\widehat{s}_{\mathrm{K}}$ is provided for K fixed. Section 2.1.2 deals with the penalized comparison to overfitting kernel selection method and a risk bound on the associated adaptive estimator.

2.1.1 Nonadaptive Risk Bounds

Let $\mathcal{K}_N$ be a set of symmetric continuous maps from $\mathbb{R}^2$ into $\mathbb{R}$, of cardinality less or equal than N, fulfilling the following assumption.

Assumption 2.1 There exists a constant $\mathfrak{m}_{\mathcal{K}} > 0$, not depending on n, such that

1. For every $\mathrm{K} \in \mathcal{K}_N$,

$$\sup_{z\in\mathbb{R}}\|\mathrm{K}(z,\cdot)\|^2 \leqslant \mathfrak{m}_{\mathcal{K}}N.$$

2. For every $\mathrm{K} \in \mathcal{K}_N$,
$$\|s_{\mathrm{K},\ell}\|^2 \leqslant \mathfrak{m}_{\mathcal{K}},$$
where
$$s_{\mathrm{K},\ell}(\cdot) = \mathbb{E}(\widehat{s}_{\mathrm{K},\ell}(\cdot)) = \mathbb{E}(\mathrm{K}(\xi, \cdot)\ell(Y)).$$

3. For every $\mathrm{K}, \mathrm{L} \in \mathcal{K}_N$,
$$\mathbb{E}(\langle \mathrm{K}(\xi_1, \cdot), \mathrm{L}(\xi_2, \cdot)\ell(Y_2)\rangle^2) \leqslant \mathfrak{m}_{\mathcal{K}}\overline{s}_{\mathrm{L},\ell},$$
where
$$\overline{s}_{\mathrm{L},\ell} = \mathbb{E}(\|\mathrm{L}(\xi, \cdot)\ell(Y)\|^2).$$

4. For every $\mathrm{K} \in \mathcal{K}_N$ and $\varphi \in \mathbb{L}^2(\mathbb{R}, dx)$,
$$\mathbb{E}(\langle \mathrm{K}(\xi, \cdot), \varphi\rangle^2) \leqslant \mathfrak{m}_{\mathcal{K}}\|\varphi\|^2.$$

In the sequel, the elements of $\mathcal{K}_N$ are called kernels and, for the sake of simplicity, $s_{\mathrm{K},\ell}$ (resp. $\overline{s}_{\mathrm{K},\ell}$) is almost always denoted by s_{K} (resp. $\overline{s}_{\mathrm{K}}$). The two following propositions provide examples of kernels sets.

Proposition 2.1 *Consider*
$$\mathcal{K}_K(h_{\min}) = \{(z, x) \mapsto K_h(z - x); h \in \mathcal{H}(h_{\min})\},$$
where $h_{\min} \in [N^{-1}, 1]$ and $\mathcal{H}(h_{\min})$ is a finite subset of $[h_{\min}, 1]$. The kernels set $\mathcal{K}_K(h_{\min})$ fulfills Assumption 2.1 and, for every $\mathrm{K} \in \mathcal{K}_K(h_{\min})$,
$$\overline{s}_{\mathrm{K}} = \frac{\|\mathrm{K}\|^2\mathbb{E}(\ell(Y)^2)}{h}.$$

See Halconruy and Marie [3], Proposition 2.2 for a proof.

Proposition 2.2 *Consider*
$$\mathcal{K}_{\mathcal{B}_N}(m_{\max}) = \left\{(z, x) \mapsto \sum_{j=1}^{m} \varphi_j(z)\varphi_j(x); m \in \{1, \ldots, m_{\max}\}\right\},$$
where $m_{\max} \in \{1, \ldots, N\}$ and $\mathcal{B}_N = (\varphi_1, \ldots, \varphi_N)$ is an I-supported orthonormal family of $\mathbb{L}^2(\mathbb{R}, dx)$. If there exists a constant $\mathfrak{c}_\varphi > 0$, not depending on N, such that
$$\sup_{x\in I} \sum_{j=1}^{m} \varphi_j(x)^2 \leqslant \mathfrak{c}_\varphi^2 m; \forall m \in \{1, \ldots, N\},$$

then the kernels set $\mathcal{K}_{\mathcal{B}_N}(m_{\max})$ *fulfills Assumption 2.1 and, for any* $\mathrm{K} \in \mathcal{K}_{\mathcal{B}_N}(m_{\max})$,

$$\overline{s}_{\mathrm{K}} \leqslant \mathfrak{c}_{\varphi}^2 \mathbb{E}(\ell(Y)^2) m.$$

See Halconruy and Marie [3], Proposition 2.3 for a proof.

In particular, a risk bound on $\widehat{s}_{\mathrm{K}}$ can be derived from the following theorem, which is the main result of this section. Theorem 2.1 is also crucial in order to establish a risk bound on the PCO (adaptive) estimator in Sect. 2.1.2.

Theorem 2.1 *Under Assumption 2.1, if*

$$\exists \alpha > 0 : \mathbb{E}(e^{\alpha|\ell(Y)|}) < \infty, \tag{2.3}$$

then there exists a constant $\mathfrak{c}_{2.1} > 0$, *not depending on* N, *such that for every* $\theta \in (0, 1)$,

$$\mathbb{E}\left(\sup_{K \in \mathcal{K}_N} \left\{ \|\widehat{s}_K - s\|^2 - (1+\theta) \right\} \left(\|s_K - s\|^2 + \frac{\overline{s}_K}{N} \right) \right] \leqslant \mathfrak{c}_{2.1} \frac{\log(N)^5}{\theta N}.$$

Sketch of proof. For a detailed proof of Theorem 2.1, see Halconruy and Marie [3], Theorem 2.8.

Step 1. For every $\mathrm{K} \in \mathcal{K}_N$,

$$\|\widehat{s}_{\mathrm{K}} - s_{\mathrm{K}}\|^2 = \frac{U_{\mathrm{K},\mathrm{K}}}{N^2} + \frac{V_{\mathrm{K}}}{N}, \tag{2.4}$$

where

$$U_{\mathrm{K},\mathrm{L}} = \sum_{i \neq k} \langle \mathrm{K}(\xi_i, \cdot)\ell(Y_i) - s_{\mathrm{K}}, \mathrm{L}(\xi_k, \cdot)\ell(Y_k) - s_{\mathrm{L}} \rangle; \forall \mathrm{K}, \mathrm{L} \in \mathcal{K}_N$$

and

$$V_{\mathrm{K}} = \frac{1}{N} \sum_{i=1}^{N} \|\mathrm{K}(\xi_i, \cdot)\ell(Y_i) - s_{\mathrm{K}}\|^2; \forall \mathrm{K} \in \mathcal{K}_N.$$

On the one hand, thanks to the concentration inequality for U-statistics of Houdré and Reynaud-Bouret [4], there exists a constant $\mathfrak{c}_1 > 0$, not depending on N, such that for every $\theta \in (0, 1)$,

$$\mathbb{E}\left(\sup_{\mathrm{K},\mathrm{L} \in \mathcal{K}_N} \left\{ \frac{|U_{\mathrm{K},\mathrm{L}}|}{N^2} - \frac{\theta}{2N} \overline{s}_{\mathrm{L}} \right\}\right) \leqslant \mathfrak{c}_1 \frac{\log(N)^5}{\theta N}. \tag{2.5}$$

On the other hand, thanks to Bernstein's inequality, there exists a constant $\mathfrak{c}_2 > 0$, not depending on N, such that for every $\theta \in (0, 1)$,

$$\mathbb{E}\left[\sup_{K\in\mathcal{K}_N}\left\{\frac{V_K}{N}-\frac{1}{N}\left(1+\frac{\theta}{2}\right)\overline{s}_K\right\}\right]\leqslant \mathfrak{c}_2\frac{\log(N)^3}{\theta N}. \tag{2.6}$$

Therefore, by Equality (2.4) together with Inequalities (2.5) and (2.6), there exists a constant $\mathfrak{c}_3 > 0$, not depending on N, such that for every $\theta \in (0, 1)$,

$$\mathbb{E}\left(\sup_{K\in\mathcal{K}_N}\left\{\|\widehat{s}_K - s_K\|^2 - \frac{1+\theta}{N}\overline{s}_K\right\}\right)\leqslant \mathfrak{c}_3\frac{\log(N)^5}{\theta N}. \tag{2.7}$$

Step 2. For every $K \in \mathcal{K}_N$ and $\theta \in (0, 1)$,

$$\begin{aligned}&\|\widehat{s}_K - s\|^2 - (1+\theta)\left(\|s_K - s\|^2 + \frac{\overline{s}_K}{N}\right)\\ &= \|\widehat{s}_K - s_K\|^2 - \frac{1+\theta}{N}\overline{s}_K + 2W_{K,K} - \theta\|s_K - s\|^2,\end{aligned} \tag{2.8}$$

where

$$W_{K,L} = \langle \widehat{s}_K - s_K, s_L - s\rangle;\ \forall K, L \in \mathcal{K}_N.$$

Thanks to Bernstein's inequality, there exists a constant $\mathfrak{c}_4 > 0$, not depending on N, such that for every $\theta \in (0, 1)$,

$$\mathbb{E}\left(\sup_{K,L\in\mathcal{K}_N}\{|2W_{K,L}| - \theta\|s_L - s\|^2\}\right)\leqslant \mathfrak{c}_4\frac{\log(N)^4}{\theta N}. \tag{2.9}$$

The conclusion comes from Equality (2.8) together with Inequalities (2.7) and (2.9).

Let us conclude this section with some remarks about Theorem 2.1:

- As mentioned above, a risk bound on the estimator $\widehat{s}_K$ can be derived from Theorem 2.1: for every $K \in \mathcal{K}_N$ and $\theta \in (0, 1)$,

$$\underbrace{\mathbb{E}(\|\widehat{s}_K - s\|^2)}_{\text{Integrated risk}} \leqslant (1+\theta)\underbrace{\|s_K - s\|^2}_{\text{Squared bias}} + \underbrace{(1+\theta)\overline{s}_K N^{-1}}_{\sim\text{Variance}} + \underbrace{\mathfrak{c}_{2.1}\theta^{-1}\log(N)^5 N^{-1}}_{\text{Negligible remainder}}.$$

  For instance, if K is of type (2.2), then

$$\|s_K - s\|^2 = \min_{\tau\in\mathcal{S}_m}\|\tau - s\|^2$$

  and $\overline{s}_K$ is of order m by Proposition 2.2, leading to

$$\mathbb{E}(\|\widehat{s}_K - s\|^2) \lesssim \min_{\tau\in\mathcal{S}_m}\|\tau - s\|^2 + \frac{m}{N}\quad\text{for}\quad m \in \{\log(N)^5, \dots, N\}.$$

- Assume that $\ell = 1$ and K is of type (2.1):

$$\mathrm{K}(z,x) = K_h(z-x);\ \forall (z,x) \in \mathbb{R}^2.$$

First, $\overline{s}_{\mathrm{K}}$ is of order $1/h$ by Proposition 2.1. Now, assume that $x \mapsto x^2K(x)$ belongs to $\mathbb{L}^1(\mathbb{R}, dx)$, and that $\mathtt{f}$ fulfills the Nikol'skii condition: there exists a constant $\mathfrak{c}_1 > 0$ such that, for every $\theta \in \mathbb{R}$,

$$\int_{-\infty}^{\infty} (\mathtt{f}(\theta + x) - \mathtt{f}(x))^2 dx \leqslant \mathfrak{c}_1 \theta^2.$$

Then, by Jensen's inequality,

$$\begin{aligned}\|s_{\mathrm{K}} - s\|^2 = \|K_h * \mathtt{f} - \mathtt{f}\|^2 &= \int_{-\infty}^{\infty} \left(\int_{-\infty}^{\infty} K(z)(\mathtt{f}(hz + x) - \mathtt{f}(x))dz \right)^2 dx \\ &\leqslant \int_{-\infty}^{\infty} K(z) \int_{-\infty}^{\infty} (\mathtt{f}(hz + x) - \mathtt{f}(x))^2 dx dz \leqslant \mathfrak{c}_1 h^2 \int_{-\infty}^{\infty} z^2 K(z) dz.\end{aligned}$$

So, by Theorem 2.1, there exists a constant $\mathfrak{c}_2 > 0$, not depending on N, such that

$$\|\widehat{s}_{\mathrm{K}} - \mathtt{f}\|^2 \leqslant \mathfrak{c}_2 \left(h^2 + \frac{1}{Nh} \right) \quad \text{for} \quad h \in \left[\frac{1}{N}, \frac{1}{\log(N)^5} \right].$$

Therefore, the bias-variance tradeoff is reached by (the risk bound on) $\widehat{s}_{\mathrm{K}}$ when h is of order $N^{-1/3}$, leading to a rate of order $N^{-2/3}$. However, in practice, h needs to be selected from data because $\mathfrak{c}_1$ depends on $\mathtt{f}$ (which is unknown).
- In Theorem 2.1, the condition (2.3) may appear too demanding. However, Y satisfies (2.3) in the density estimation framework (i.e., when $\ell = 1$), or in the nonparametric regression framework when Y has a compactly supported distribution or a Gaussian one (by Fernique's theorem).

2.1.2 *Kernel Selection: The Penalized Comparison to Overfitting Method*

This section deals with a risk bound on the (adaptive) PCO estimator $\widehat{s}_{\widehat{\mathrm{K}}}$, where

$$\widehat{\mathrm{K}} = \arg\min_{\mathrm{K} \in \mathcal{K}_N} \{\|\widehat{s}_{\mathrm{K}} - \widehat{s}_{\mathrm{K}_0}\|^2 + \mathrm{pen}_\ell(\mathrm{K})\}, \tag{2.10}$$

with K_0 an overfitting kernel in the sense that

$$\mathrm{K}_0 = \arg\max_{\mathrm{K} \in \mathcal{K}_N} \left\{ \sup_{x \in \mathbb{R}} |\mathrm{K}(x,x)| \right\},$$

and

$$\text{pen}_\ell(\text{K}) = \frac{2}{N^2}\sum_{i=1}^{N}\langle \text{K}(\xi_i,\cdot), \text{K}_0(\xi_i,\cdot)\rangle \ell(Y_i)^2;\ \forall \text{K} \in \mathcal{K}_N.$$

In the PCO (kernel selection) criterion (2.10), the *overfitting* loss $\text{K} \mapsto \|\widehat{s}_\text{K} - \widehat{s}_{\text{K}_0}\|^2$, which models the risk to select $\text{K} \in \mathcal{K}_N$ too close to K_0, and then to excessively degrade the variance of $\widehat{s}_\text{K}$, is penalized by $\text{pen}_\ell(\text{K})$ which is of same order as the variance term in Theorem 2.1. The PCO method has been introduced by C. Lacour, P. Massart and V. Rivoirard in [1] for $\ell = 1$ and $\mathcal{K}_N = \mathcal{K}_K(h_{\min})$ (see Proposition 2.1). For this kernels set,

$$\text{K}_0(z, x) = K_{h_{\min}}(z - x);\ \forall x, z \in \mathbb{R},$$

and for $\mathcal{K}_N = \mathcal{K}_{\mathcal{B}_N}(m_{\max})$ (see Proposition 2.2),

$$\text{K}_0(z, x) = \sum_{j=1}^{m_{\max}} \varphi_j(z)\varphi_j(x);\ \forall x, z \in \mathbb{R}.$$

In the sequel, in addition to Assumption 2.1, the kernels set $\mathcal{K}_N$ fulfills one of the two following assumptions.

Assumption 2.2 There exists a constant $\overline{\mathfrak{m}}_\mathcal{K} > 0$, not depending on N, such that

$$\mathbb{E}\left(\sup_{\text{K},\text{L}\in\mathcal{K}_N} \langle \text{K}(\xi,\cdot), s_\text{L}\rangle^2\right) \leqslant \overline{\mathfrak{m}}_\mathcal{K}.$$

Assumption 2.3 The function s is bounded, and

$$\widetilde{\mathfrak{m}}_\mathcal{K} = \sup\{\|\text{K}(z,\cdot)\|_1^2;\ \text{K} \in \mathcal{K}_N\ ,\ z \in \mathbb{R}\} \quad \text{doesn't depend on } N.$$

Under Assumption 2.3, $\mathcal{K}_N$ fulfills Assumption 2.2. Indeed,

$$\begin{aligned}
&\mathbb{E}\left(\sup_{\text{K},\text{L}\in\mathcal{K}_N} \langle \text{K}(\xi,\cdot), s_\text{L}\rangle^2\right)\\
&\qquad \leqslant \left(\sup_{\text{L}\in\mathcal{K}_N} \|s_\text{L}\|_\infty^2\right)\mathbb{E}\left(\sup_{\text{K}\in\mathcal{K}_N} \|\text{K}(\xi,\cdot)\|_1^2\right)\\
&\qquad \leqslant \widetilde{\mathfrak{m}}_\mathcal{K} \sup\left\{\left(\int_{-\infty}^{\infty} |\text{L}(z,x)s(x)|dx\right)^2;\ \ \text{L} \in \mathcal{K}_N, z \in \mathbb{R}\right\} \leqslant \widetilde{\mathfrak{m}}_\mathcal{K}^2\|s\|_\infty^2.
\end{aligned}$$

Let us make some remarks about Assumptions 2.2 and 2.3:

- For instance, $s = b_0\text{f}$ is bounded in the linear regression model with Gaussian inputs. So, to assume that s is bounded is not that demanding.

- For every $\mathtt{K} \in \mathcal{K}_K(h_{\min})$, there exists $h \in \mathcal{H}(h_{\min})$ such that

$$\mathtt{K}(z, x) = K_h(z - x); \forall x, z \in \mathbb{R},$$

  and then $\|\mathtt{K}(z, \cdot)\|_1 = \|K\|_1$ for every $z \in \mathbb{R}$. So, the kernels set $\mathcal{K}_K(h_{\min})$ fulfills Assumption 2.3 with $\widetilde{\mathfrak{m}}_{\mathcal{K}} = \|K\|_1^2$.
- Consider $\mathtt{l}, \mathtt{r} \in \mathbb{R}$ satisfying $\mathtt{l} < \mathtt{r}$, and assume that $\mathcal{B}_N = (\varphi_1, \dots, \varphi_N)$ is the $[\mathtt{l}, \mathtt{r}]$-supported trigonometric basis such that, for every $x \in \mathbb{R}$ and $j \in \mathbb{N}^*$ satisfying $2j + 1 \leqslant N$,

$$\varphi_1(x) = \sqrt{\frac{1}{\mathtt{r} - \mathtt{l}}}\mathbf{1}_{[\mathtt{l},\mathtt{r}]}(x),$$
$$\varphi_{2j+1}(x) = \sqrt{\frac{2}{\mathtt{r} - \mathtt{l}}} \sin\left(2\pi j \frac{x - \mathtt{l}}{\mathtt{r} - \mathtt{l}}\right)\mathbf{1}_{[\mathtt{l},\mathtt{r}]}(x) \text{ and}$$
$$\varphi_{2j}(x) = \sqrt{\frac{2}{\mathtt{r} - \mathtt{l}}} \cos\left(2\pi j \frac{x - \mathtt{l}}{\mathtt{r} - \mathtt{l}}\right)\mathbf{1}_{[\mathtt{l},\mathtt{r}]}(x).$$

  Then, the kernels set $\mathcal{K}_{\mathcal{B}_N}(m_{\max})$ fulfills Assumption 2.2 but not Assumption 2.3. See Halconruy and Marie [3], Proposition 3.6 for a proof.

The following theorem, which is the main result of this section, provides a risk bound on the PCO estimator $\widehat{s}_{\widehat{\mathtt{K}}}$.

Theorem 2.2 *Under Assumptions 2.1 and 2.2, if Y satisfies (2.3), then there exists a constant $\mathfrak{c}_{2.2} > 0$, not depending on N, such that for every $\theta \in (0, 1)$,*

$$\mathbb{E}(\|\widehat{s}_{\widehat{\mathtt{K}}} - s\|^2) \leqslant (1 + \theta) \min_{\mathtt{K} \in \mathcal{K}_N} \mathbb{E}(\|\widehat{s}_{\mathtt{K}} - s\|^2) + \frac{\mathfrak{c}_{2.2}}{\theta}\left(\|s_{K_0} - s\|^2 + \frac{\log(N)^5}{N}\right).$$

Sketch of proof. For a detailed proof of Theorem 2.2, see Halconruy and Marie [3], Theorem 3.2

Step 1. By following the same line as in the proof of Theorem 2.1, there exists a constant $\mathfrak{c}_1 > 0$, not depending on N, such that for every $\theta \in (0, 1)$,

$$\mathbb{E}\left(\sup_{\mathtt{K} \in \mathcal{K}_N}\left\{\|s_{\mathtt{K}} - s\|^2 + \frac{\overline{s}_{\mathtt{K}}}{N} - \frac{1}{1 - \theta}\|\widehat{s}_{\mathtt{K}} - s\|^2\right\}\right) \leqslant \mathfrak{c}_1 \frac{\log(N)^5}{\theta(1 - \theta)N}.$$

This inequality together with Theorem 2.1 provide a two-sided relationship between

$$\|\widehat{s}_{\mathtt{K}} - s\|^2 \quad \text{and} \quad \|s_{\mathtt{K}} - s\|^2; \mathtt{K} \in \mathcal{K}_N.$$

Step 2. First,

$$\|\widehat{s}_{\widehat{K}} - s\|^2 = \|\widehat{s}_{\widehat{K}} - \widehat{s}_{K_0}\|^2 + \|\widehat{s}_{K_0} - s\|^2 + 2\langle \widehat{s}_{\widehat{K}} - \widehat{s}_{K_0}, \widehat{s}_{K_0} - s\rangle,$$

and, for any $K \in \mathcal{K}_N$,

$$\begin{aligned}\|\widehat{s}_{\widehat{K}} - \widehat{s}_{K_0}\|^2 &\leqslant \|\widehat{s}_K - \widehat{s}_{K_0}\|^2 + \text{pen}_\ell(K) - \text{pen}_\ell(\widehat{K}) \quad \text{by (2.10)}\\ &= \|\widehat{s}_K - s\|^2 + 2\langle \widehat{s}_K - s, s - \widehat{s}_{K_0}\rangle + \|s - \widehat{s}_{K_0}\|^2 + \text{pen}_\ell(K) - \text{pen}_\ell(\widehat{K})\\ &= \|\widehat{s}_K - s\|^2 + 2\langle \widehat{s}_K - \widehat{s}_{K_0}, s - \widehat{s}_{K_0}\rangle - \|s - \widehat{s}_{K_0}\|^2 + \text{pen}_\ell(K) - \text{pen}_\ell(\widehat{K}).\end{aligned}$$

Then,

$$\begin{aligned}\|\widehat{s}_{\widehat{K}} - s\|^2 &\leqslant \|\widehat{s}_K - s\|^2 + 2\langle \widehat{s}_K - \widehat{s}_{K_0}, s - \widehat{s}_{K_0}\rangle\\ &\qquad + \text{pen}_\ell(K) - \text{pen}_\ell(\widehat{K}) + 2\langle \widehat{s}_{\widehat{K}} - \widehat{s}_{K_0}, \widehat{s}_{K_0} - s\rangle\\ &= \|\widehat{s}_K - s\|^2 + \text{pen}_\ell(K) - \text{pen}_\ell(\widehat{K}) + 2\langle \widehat{s}_{\widehat{K}} - \widehat{s}_K, \widehat{s}_{K_0} - s\rangle\\ &= \|\widehat{s}_K - s\|^2 - \psi(K) + \psi(\widehat{K}),\end{aligned} \tag{2.11}$$

where

$$\psi(\cdot) = 2\langle \widehat{s}_\cdot - s, \widehat{s}_{K_0} - s\rangle - \text{pen}_\ell(\cdot).$$

Now, let us rewrite $\psi(\cdot)$ in terms of $U_{\cdot,K_0}$, $W_{\cdot,K_0}$ and $W_{K_0,\cdot}$. For any $K \in \mathcal{K}_N$,

$$\begin{aligned}\psi(K) &= 2\langle \widehat{s}_K - s_K + s_K - s, \widehat{s}_{K_0} - s_{K_0} + s_{K_0} - s\rangle - \text{pen}_\ell(K)\\ &= 2\langle \widehat{s}_K - s_K, \widehat{s}_{K_0} - s_{K_0}\rangle - \text{pen}_\ell(K)\\ &\qquad + 2\underbrace{(W_{K,K_0} + W_{K_0,K} + \langle s_K - s, s_{K_0} - s\rangle)}_{=:\psi_3(K)}\\ &= 2\psi_3(K) + \frac{2U_{K,K_0}}{N^2}\\ &\qquad \underbrace{+ 2\left(\frac{1}{N^2}\sum_{i=1}^{N}\langle K(\xi_i,\cdot)\ell(Y_i) - s_K, K_0(\xi_i,\cdot)\ell(Y_i) - s_{K_0}\rangle - \frac{\text{pen}_\ell(K)}{2}\right)}_{=:\psi_2(K)}\end{aligned}$$

and, by the definition of $\text{pen}_\ell(K)$,

$$\psi_2(K) = -\frac{1}{N^2}\left(\sum_{i=1}^{N}\langle K(\xi^i,\cdot), s_{K_0}\rangle\ell(Y_i) + \sum_{i=1}^{N}\langle K_0(\xi^i,\cdot), s_K\rangle\ell(Y_i)\right) + \frac{1}{N}\langle s_K, s_{K_0}\rangle.$$

So,

$$\psi(K) = 2(\psi_1(K) + \psi_2(K) + \psi_3(K)) \quad \text{with} \quad \psi_1(K) = \frac{U_{K,K_0}}{N^2}.$$

Step 3. Consider $\texttt{K} \in \mathcal{K}_N$ and $\mathfrak{K} = \texttt{K}$ or $\widehat{\texttt{K}}$. By controlling $\mathbb{E}(|\psi_1(\mathfrak{K})|)$ thanks to Inequality (2.5), $\mathbb{E}(|\psi_2(\mathfrak{K})|)$ thanks to Assumptions 2.1.(2) and 2.2, and $\mathbb{E}(|\psi_3(\mathfrak{K})|)$ thanks to Inequality (2.9), there exists a constant $\mathfrak{c}_2 > 0$, not depending on N and $\mathfrak{K}$, such that for any $\theta \in (0, 1)$,

$$\mathbb{E}(|\psi(\mathfrak{K})|) \leqslant \theta\mathbb{E}\left(\|s_{\mathfrak{K}} - s\|^2 + \frac{\overline{s}_{\mathfrak{K}}}{N}\right) + \left(\frac{\theta}{2} + \frac{2}{\theta}\right)\|s_{\texttt{K}_0} - s\|^2 + \mathfrak{c}_2\frac{\log(N)^5}{\theta N}.$$

So, by Step 1,

$$\mathbb{E}(|\psi(\mathfrak{K})|) \leqslant \frac{\theta}{1-\theta}\mathbb{E}(\|\widehat{s}_{\mathfrak{K}} - s\|^2) + \left(\frac{\theta}{2} + \frac{2}{\theta}\right)\|s_{\texttt{K}_0} - s\|^2 + \left(\frac{\mathfrak{c}_1}{1-\theta} + \frac{\mathfrak{c}_2}{\theta}\right)\frac{\log(N)^5}{N}. \tag{2.12}$$

By Inequalities (2.11) and (2.12), there exists a constant $\mathfrak{c}_3 > 0$, not depending on N, $\texttt{K}$ and θ, such that

$$\left(1 - \frac{\theta}{1-\theta}\right)\mathbb{E}(\|\widehat{s}_{\widehat{\texttt{K}}} - s\|^2) \leqslant \left(1 + \frac{\theta}{1-\theta}\right)\mathbb{E}(\|\widehat{s}_{\texttt{K}} - s\|^2) + \frac{\mathfrak{c}_3}{\theta}\left(\|s_{\texttt{K}_0} - s\|^2 + \frac{\log(N)^5}{(1-\theta)N}\right)$$

and, by taking $\theta \in (0, 1/2)$, the conclusion comes from Theorem 2.1.

Notation. Consider $\texttt{l}, \texttt{r} \in \mathbb{R}$ such that $\texttt{l} < \texttt{r}$. For every $\varphi \in \mathbb{L}^2(\mathbb{R}, \texttt{f}(x)dx)$,

$$\|\varphi\|_{\texttt{f},\texttt{l},\texttt{r}} = \left(\int_{\texttt{l}}^{\texttt{r}} \varphi(x)^2\texttt{f}(x)dx\right)^{\frac{1}{2}} ; \forall\varphi \in \mathbb{L}^2(\mathbb{R}, \texttt{f}(x)dx).$$

Now, assume that $\texttt{f}(x) > m$ for every $x \in [\texttt{l}, \texttt{r}]$ ($m \in (0, 1]$). Assume also that $\ell(y) = y$ for every $y \in \mathbb{R}$, and that $(\xi_1, Y_1), \dots, (\xi_N, Y_N)$ satisfy (1.5). The following corollary provides a risk bound on the adaptive 2-kernels Nadaraya-Watson estimator

$$\widehat{b}_{\widehat{\texttt{K}},\widehat{\texttt{L}}}(x) = \frac{\widehat{s}_{\widehat{\texttt{K}},\ell}(x)}{\widehat{s}_{\widehat{\texttt{L}},1}(x)}\mathbf{1}_{\widehat{s}_{\widehat{\texttt{L}},1}(x) > \frac{m}{2}} ; x \in [\texttt{l}, \texttt{r}],$$

where $\widehat{\texttt{K}}$ and $\widehat{\texttt{L}}$ are both selected in $\mathcal{K}_N$ via the PCO method.

Corollary 2.1 *Under Assumptions 2.1 and 2.2 for both ℓ and 1, if $\texttt{f}$ and $b_0^2\texttt{f}$ are bounded, and if Y satisfies (2.3), then there exists a constant $\mathfrak{c}_{2.1} > 0$, not depending on N, $\texttt{l}$ and $\texttt{r}$, such that for every $\theta \in (0, 1)$,*

$$\mathbb{E}(\|\widehat{b}_{\widehat{\mathtt{K}},\widehat{\mathtt{L}}} - b_0\|^2_{\mathtt{f},1,\mathtt{r}}) \leqslant \frac{\mathfrak{c}_{2.1}}{m^2}[(1+\theta) \min_{(\mathtt{K},\mathtt{L})\in\mathcal{K}_N^2} \{\mathbb{E}(\|\widehat{s}_{\mathtt{K},\ell} - b_0\mathtt{f}\|^2) + \mathbb{E}(\|\widehat{s}_{\mathtt{L},1} - \mathtt{f}\|^2)\}$$

$$+ \frac{1}{\theta}\left(\|s_{\mathtt{K}_0,\ell} - b_0\mathtt{f}\|^2 + \|s_{\mathtt{K}_0,1} - \mathtt{f}\|^2 + \frac{\log(N)^5}{N}\right)],$$

where

$$\widehat{\mathtt{K}} = \arg\min_{\mathtt{K}\in\mathcal{K}_N}\{\|\widehat{s}_{\mathtt{K},\ell} - \widehat{s}_{\mathtt{K}_0,\ell}\|^2 + \mathrm{pen}_\ell(\mathtt{K})\}$$

$$\text{and} \quad \widehat{\mathtt{L}} = \arg\min_{\mathtt{L}\in\mathcal{K}_N}\{\|\widehat{s}_{\mathtt{L},1} - \widehat{s}_{\mathtt{K}_0,1}\|^2 + \mathrm{pen}_1(\mathtt{L})\}.$$

Proof First of all,

$$\widehat{b}_{\widehat{\mathtt{K}},\widehat{\mathtt{L}}} - b_0 = \left(\frac{\widehat{s}_{\widehat{\mathtt{K}},\ell} - b_0\mathtt{f}}{\widehat{s}_{\widehat{\mathtt{L}},1}} + \left(\frac{1}{\widehat{s}_{\widehat{\mathtt{L}},1}} - \frac{1}{\mathtt{f}}\right) b_0\mathtt{f}\right)\mathbf{1}_{\widehat{s}_{\widehat{\mathtt{L}},1}(\cdot)>\frac{m}{2}} - b_0\mathbf{1}_{\widehat{s}_{\widehat{\mathtt{L}},1}(\cdot)\leqslant\frac{m}{2}}.$$

Then,

$$\|\widehat{b}_{\widehat{\mathtt{K}},\widehat{\mathtt{L}}} - b_0\|^2_{\mathtt{f},1,\mathtt{r}} = \|b_0\mathbf{1}_{\widehat{s}_{\widehat{\mathtt{L}},1}}(\cdot)\leqslant\frac{m}{2}\|^2_{\mathtt{f},1,\mathtt{r}}$$

$$+ \left\|\left(\frac{\widehat{s}_{\widehat{\mathtt{K}},\ell} - b_0\mathtt{f}}{\widehat{s}_{\widehat{\mathtt{L}},1}} + \left(\frac{1}{\widehat{s}_{\widehat{\mathtt{L}},1}} - \frac{1}{\mathtt{f}}\right) b_0\mathtt{f}\right)\mathbf{1}_{\widehat{s}_{\widehat{\mathtt{L}},1}(\cdot)>\frac{m}{2}}\right\|^2_{\mathtt{f},1,\mathtt{r}}.$$

Moreover, for any $x \in [1, \mathtt{r}]$, since $\mathtt{f}(x) > m$, for every $\omega \in \{\widehat{s}_{\widehat{\mathtt{L}},1}(\cdot) \leqslant m/2\}$,

$$|\mathtt{f}(x) - \widehat{s}_{\widehat{\mathtt{L}},1}(x,\omega)| \geqslant \mathtt{f}(x) - \widehat{s}_{\widehat{\mathtt{L}},1}(x,\omega) > m - \frac{m}{2} = \frac{m}{2}.$$

Thus,

$$\|\widehat{b}_{\widehat{\mathtt{K}},\widehat{\mathtt{L}}} - b_0\|^2_{\mathtt{f},1,\mathtt{r}} \leqslant \frac{8}{m^2}\|\widehat{s}_{\widehat{\mathtt{K}},\ell} - b_0\mathtt{f}\|^2_{\mathtt{f},1,\mathtt{r}}$$

$$+ \frac{8}{m^2}\|(\mathtt{f} - \widehat{s}_{\widehat{\mathtt{L}},1})b_0\|^2_{\mathtt{f},1,\mathtt{r}} + 2\|b_0\mathbf{1}_{|\mathtt{f}(\cdot)-\widehat{s}_{\widehat{\mathtt{L}},1}(\cdot)|>\frac{m}{2}}\|^2_{\mathtt{f},1,\mathtt{r}}$$

$$\leqslant \frac{8}{m^2}\int_{-\infty}^{\infty}(\widehat{s}_{\widehat{\mathtt{K}},\ell} - b_0\mathtt{f})(x)^2\mathtt{f}(x)dx$$

$$+ \frac{8}{m^2}\int_1^{\mathtt{r}}(\mathtt{f}(x) - \widehat{s}_{\widehat{\mathtt{L}},1}(x))^2 b_0(x)^2\mathtt{f}(x)dx$$

$$+ 2\int_1^{\mathtt{r}} b_0(x)^2\mathtt{f}(x)\mathbf{1}_{|\mathtt{f}(x)-\widehat{s}_{\widehat{\mathtt{L}},1}(x)|>\frac{m}{2}}dx.$$

Since $\texttt{f}$ and $b_0^2\texttt{f}$ are bounded on $\mathbb{R}$,

$$\|\widehat{b}_{\widehat{\texttt{K}},\widehat{\texttt{L}}} - b_0\|_{\texttt{f},1,\texttt{r}}^2 \leqslant \frac{8\|\texttt{f}\|_\infty}{m^2}\|\widehat{s}_{\widehat{\texttt{K}},\ell} - b_0\texttt{f}\|^2 + \frac{8\|b_0^2\texttt{f}\|_\infty}{m^2}\|\widehat{s}_{\widehat{\texttt{L}},1} - \texttt{f}\|^2 + 2\|b_0^2\texttt{f}\|_\infty\int_{-\infty}^{\infty}\mathbf{1}_{|\texttt{f}(x)-\widehat{s}_{\widehat{\texttt{L}},1}(x)|>\frac{m}{2}}dx.$$

Therefore, by Markov's inequality,

$$\begin{aligned}\mathbb{E}(\|\widehat{b}_{\widehat{\texttt{K}},\widehat{\texttt{L}}} - b_0\|_{\texttt{f},1,\texttt{r}}^2) &\leqslant \frac{8\|\texttt{f}\|_\infty}{m^2}\mathbb{E}(\|\widehat{s}_{\widehat{\texttt{K}},\ell} - b_0\texttt{f}\|^2) + \frac{8\|b_0^2\texttt{f}\|_\infty}{m^2}\mathbb{E}(\|\widehat{s}_{\widehat{\texttt{L}},1} - \texttt{f}\|^2)\\ &\quad + \frac{8\|b_0^2\texttt{f}\|_\infty}{m^2}\int_{-\infty}^{\infty}\mathbb{E}((\texttt{f}(x) - \widehat{s}_{\widehat{\texttt{L}},1}(x))^2)dx\\ &\leqslant \frac{8(\|\texttt{f}\|_\infty \vee \|b_0^2\texttt{f}\|_\infty)}{m^2}(\mathbb{E}(\|\widehat{s}_{\widehat{\texttt{K}},\ell} - b_0\texttt{f}\|^2) + 2\mathbb{E}(\|\widehat{s}_{\widehat{\texttt{L}},1} - \texttt{f}\|^2)).\end{aligned}$$

Theorem 2.2 allows to conclude. □

Let us conclude this section with some remarks about Corollary 2.1:

- By Corollary 2.1, the risk of $\widehat{b}_{\widehat{\texttt{K}},\widehat{\texttt{L}}}$ is controlled by the sum of the minimal risks of

$$\widehat{s}_{\texttt{K},\ell} \text{ and } \widehat{s}_{\texttt{L},1};\ (\texttt{K},\texttt{L}) \in \mathcal{K}_N^2,$$

  up to a multiplicative constant.
- The limitation of Corollary 2.1 is that m is unknown in practice. Then, it needs to be estimated, for instance by

$$\widehat{m}_{\widehat{\texttt{L}}} = \min\{\widehat{s}_{\widehat{\texttt{L}},1}(x);\ x \in [\texttt{l},\texttt{r}]\}.$$

2.2 The Projection Least Squares Estimator of the Regression Function

This section deals with a nonadaptive risk bound on the projection least squares estimator $\widehat{b}_m$ of b_0 computed from $(\xi_1, Y_1), \dots, (\xi_N, Y_N)$, and then with model selection. Recall that $\widehat{b}_m$ is a minimizer of the objective function $\gamma_N^{\texttt{r}}$ defined by

$$\gamma_N^{\texttt{r}}(b) = \frac{1}{N}\sum_{i=1}^{N}(b(\xi_i)^2 - 2b(\xi_i)Y_i);\ \forall b \in \mathcal{S}_m.$$

First of all, for $b = \sum_{j=1}^{m} \theta_j \varphi_j$ with $\theta_1, \dots, \theta_m \in \mathbb{R}$,

$$\nabla \gamma_N^{\mathfrak{r}}(b) = 2(\widehat{G}_m \theta - \widehat{V}_m),$$

where $\theta = (\theta_1, \dots, \theta_m)$, $\widehat{G}_m = (\langle \varphi_j, \varphi_\ell \rangle_{\mathfrak{r},N})_{j,\ell}$, $\langle \cdot, \cdot \rangle_{\mathfrak{r},N}$ is the empirical inner product defined by

$$\langle \varphi, \psi \rangle_{\mathfrak{r},N} := \frac{1}{N} \sum_{i=1}^{N} \varphi(\xi_i)\psi(\xi_i),$$

and

$$\widehat{V}_m = \left(\frac{1}{N} \sum_{i=1}^{N} \varphi_j(\xi_i) Y_i \right)_{j \in \{1,\dots,m\}}.$$

The symmetric matrix $\widehat{G}_m$ is positive semidefinite because, for every $x \in \mathbb{R}^m$,

$$x^* \widehat{G}_m x = \frac{1}{N} \sum_{i=1}^{N} \left(\sum_{j=1}^{m} x_j \varphi_j(\xi_i) \right)^2 \geqslant 0.$$

If in addition $\widehat{G}_m$ is invertible, it is positive definite, and then

$$\widehat{b}_m = \sum_{j=1}^{m} \widehat{\theta}_j \varphi_j \quad \text{with} \quad \widehat{\theta} = \widehat{G}_m^{-1} \widehat{V}_m$$

is the only minimizer of $\gamma_N^{\mathfrak{r}}$ in $\mathcal{S}_m$. The following proposition provides a simple risk bound on $\widehat{b}_m$.

Proposition 2.3 *Consider $b_0^I = b_0 \mathbf{1}_I$ and σ^2 the common variance of the ε_i's. If $\widehat{G}_m$ is almost surely invertible, and if $b_0^I \in \mathbb{L}^2(\mathbb{R}, \mathfrak{f}(x)dx)$, then*

$$\mathbb{E}(\|\widehat{b}_m - b_0^I\|_{\mathfrak{r},N}^2) \leqslant \min_{\tau \in \mathcal{S}_m} \|\tau - b_0^I\|_{\mathfrak{f}}^2 + \sigma^2 \frac{m}{N}.$$

Proof Let us introduce two empirical maps:

- The empirical process ν_N, defined by

$$\nu_N(\tau) := \frac{1}{N} \sum_{i=1}^{N} \tau(\xi_i) \varepsilon_i.$$

Note that

$$[\widehat{V}_m]_j = \langle b_0, \varphi_j \rangle_{\mathfrak{r},N} + \nu_N(\varphi_j); \forall j \in \{1, \dots, m\}. \tag{2.13}$$

- The empirical orthogonal projection $\widehat{\Pi}_m$ from $\mathbb{L}^2(\mathbb{R}, \mathfrak{f}(x)dx)$ onto $\mathcal{S}_m$, defined by

$$\widehat{\Pi}_m(\cdot) \in \arg\min_{\tau \in \mathcal{S}_m} \|\tau - \cdot\|^2_{\mathfrak{r},N}. \tag{2.14}$$

Consider $h \in \mathbb{L}^2(\mathbb{R}, \mathfrak{f}(x)dx)$. For $\tau = \sum_{j=1}^m \pi_j \varphi_j$ with $\pi = (\pi_1, \dots, \pi_m) \in \mathbb{R}^m$,

$$\nabla_\tau \|\tau - h\|^2_{\mathfrak{r},N} = 2(\widehat{G}_m \pi - (\langle h, \varphi_j \rangle_{\mathfrak{r},N})_j).$$

Since $\widehat{G}_m$ is almost surely invertible, the minimization problem (2.14) has (a.s.) a unique solution:

$$\widehat{\Pi}_m(h) = \sum_{j=1}^m [\widehat{G}_m^{-1} \widehat{P}_m(h)]_j \varphi_j$$
$$\text{with} \quad \widehat{P}_m(h) = (\langle h, \varphi_j \rangle_{\mathfrak{r},N})_{j \in \{1,\dots,m\}}.$$

First, by the definition of $\widehat{\Pi}_m$ (see (2.14)),

$$\|\widehat{b}_m - b_0^I\|^2_{\mathfrak{r},N} = \min_{\tau \in \mathcal{S}_m} \|\tau - b_0^I\|^2_{\mathfrak{r},N} + \|\widehat{b}_m - \widehat{\Pi}_m(b_0^I)\|^2_{\mathfrak{r},N}.$$

Now, by the decompositions of $\widehat{b}_m$ and $\widehat{\Pi}_m(b_0^I)$ in the basis $(\varphi_1, \dots, \varphi_m)$ of $\mathcal{S}_m$, and by (2.13),

$$\begin{aligned}\widehat{b}_m(\cdot) - \widehat{\Pi}_m(b_0^I)(\cdot) &= \langle \widehat{G}_m^{-1}(\widehat{V}_m - \widehat{P}_m(b_0^I)), (\varphi_j(\cdot))_j \rangle_{2,m} \\ &= \langle \widehat{G}_m^{-1} \widehat{\Delta}_m, (\varphi_j(\cdot))_j \rangle_{2,m} \quad \text{with} \quad \widehat{\Delta}_m = (\nu_N(\varphi_j))_{j \in \{1,\dots,m\}}.\end{aligned}$$

Then,

$$\begin{aligned}\|\widehat{b}_m - \widehat{\Pi}_m(b_0^I)\|^2_{\mathfrak{r},N} &= \frac{1}{N} \sum_{i=1}^N \left(\sum_{j=1}^m [\widehat{G}_m^{-1} \widehat{\Delta}_m]_j \varphi_j(\xi_i) \right)^2 \\ &= \sum_{j,\ell=1}^m [\widehat{G}_m^{-1} \widehat{\Delta}_m]_j [\widehat{G}_m^{-1} \widehat{\Delta}_m]_\ell \langle \varphi_j, \varphi_\ell \rangle_{\mathfrak{r},N} = \widehat{\Delta}_m^* \widehat{G}_m^{-1} \widehat{\Delta}_m \quad .\end{aligned} \tag{2.15}$$

Since the ε_i's are independent of the ξ_i's,

$$\begin{aligned}
\mathbb{E}(\|\widehat{b}_m - \widehat{\Pi}_m(b_0^I)\|_{\mathfrak{r},N}^2) &= \sum_{j,\ell=1}^{m} \mathbb{E}([\widehat{\Delta}_m]_j[\widehat{\Delta}_m]_\ell[\widehat{G}_m^{-1}]_{j,\ell}) \\
&= \frac{1}{N^2}\sum_{j,\ell=1}^{m}\sum_{i,k=1}^{N} \mathbb{E}(\varphi_j(\xi_i)\varphi_\ell(\xi_k)[\widehat{G}_m^{-1}]_{j,\ell})\mathbb{E}(\varepsilon_i\varepsilon_k) \\
&= \frac{\sigma^2}{N}\sum_{j,\ell=1}^{m} \mathbb{E}(\langle\varphi_j,\varphi_\ell\rangle_{\mathfrak{r},N}[\widehat{G}_m^{-1}]_{j,\ell}) \\
&= \frac{\sigma^2}{N}\sum_{j=1}^{m} \mathbb{E}\left(\sum_{\ell=1}^{m}[\widehat{G}_m]_{j,\ell}[\widehat{G}_m^{-1}]_{\ell,j}\right) \\
&= \frac{\sigma^2}{N}\sum_{j=1}^{m} \mathbb{E}([\widehat{G}_m\widehat{G}_m^{-1}]_{j,j}) = \sigma^2\frac{m}{N}.
\end{aligned} \tag{2.16}$$

Therefore,

$$\mathbb{E}(\|\widehat{b}_m - b_0^I\|_{\mathfrak{r},N}^2) = \underbrace{\mathbb{E}\left(\min_{\tau\in S_m}\|\tau - b_0^I\|_{\mathfrak{r},N}^2\right)}_{\leqslant \min_\tau \|\tau - b_0^I\|_{\mathfrak{f}}^2} + \sigma^2\frac{m}{N}.$$

□

The order of the *bias term*

$$\min_{\tau\in S_m}\|\tau - b_0^I\|_{\mathfrak{f}}^2 \quad \text{in Proposition 2.3 depends on the } \varphi_j'\text{s.}$$

For instance, assume that $(\varphi_1, \dots, \varphi_m)$ is the I-supported trigonometric basis with $I = [0, 1]$. Consider $\beta \in \mathbb{N}^*$, the Fourier-Sobolev space

$$\mathbb{W}_2^\beta([0,1]) = \left\{\varphi : [0,1] \to \mathbb{R}\beta \text{ times differentiable} : \int_0^1 \varphi^{(\beta)}(x)^2 dx < \infty\right\},$$

and assume that $(b_0)_{|I} \in \mathbb{W}_0^\beta([0, 1])$. By DeVore and Lorentz [5], Corollary 2.4, p. 205, there exists a constant $\mathfrak{c}_\beta > 0$, not depending on m, such that

$$\|\Pi_m(b_0^I) - b_0^I\|^2 \leqslant \mathfrak{c}_\beta m^{-2\beta},$$

where Π_m is the orthogonal projection from $\mathbb{L}^2(\mathbb{R}, dx)$ onto $\mathcal{S}_m$. If in addition $\mathfrak{f}$ is upper bounded on I by a constant $\overline{\mathfrak{m}} > 0$, then

$$\min_{\tau \in \mathcal{S}_m} \|\tau - b_0^I\|_{\mathfrak{f}}^2 \leqslant \overline{\mathfrak{m}} \|\Pi_m(b_0^I) - b_0^I\|^2$$
$$\leqslant \overline{\mathfrak{c}}_\beta m^{-2\beta} \quad \text{with} \quad \overline{\mathfrak{c}}_\beta = \mathfrak{c}_\beta \overline{\mathfrak{m}}.$$

In conclusion, by Proposition 2.3,

$$\mathbb{E}(\|\widetilde{b}_m - b_0^I\|_{\mathfrak{r},N}^2) \leqslant \mathfrak{c}_{2.3}\left(m^{-2\beta} + \frac{m}{N}\right) \quad \text{with} \quad \mathfrak{c}_{2.3} = \overline{\mathfrak{c}}_\beta \vee \sigma^2,$$

and then the bias-variance tradeoff is reached by (the risk bound on) $\widetilde{b}_m$ for m of order $N^{1/(1+2\beta)}$.

Now, assume that I is compact for the sake of simplicity, and consider

$$\widehat{m} = \arg\min_{m \in \mathcal{M}_N} \{-\|\widehat{b}_m\|_{\mathfrak{r},N}^2 + \text{pen}(m)\},$$

where

$$\text{pen}(m) = \mathfrak{c}_{\text{cal}} \sigma^2 \frac{m}{N}; \ \forall m \in \{1, \dots, N\},$$

the constant $\mathfrak{c}_{\text{cal}} > 0$ needs to be calibrated in practice, and

$$\mathcal{M}_N = \left\{m \in \{1, \dots, N\} : \mathfrak{L}(m) \leqslant \frac{N}{\log(N)}\right\} \quad \text{with} \quad \mathfrak{L}(m) = 1 \vee \left(\sup_{x \in I} \sum_{j=1}^m \varphi_j(x)^2\right).$$

Note that, in fact,

$$\widehat{m} = \arg\min_{m \in \mathcal{M}_N} \{\gamma_m(\widehat{b}_m) + \text{pen}(m)\}$$

because, for every $m \in \mathcal{M}_N$,

$$\begin{aligned}
\gamma_N^{\mathfrak{r}}(\widehat{b}_m) - \|\widehat{b}_m\|_{\mathfrak{r},N}^2 &= -\frac{2}{N} \sum_{i=1}^N Y_i \sum_{j=1}^m \widehat{\theta}_j \varphi_j(\xi_i) \\
&= -2 \sum_{j=1}^m \widehat{\theta}_j [\widehat{V}_m]_j = -2\widehat{V}_m^* \widehat{\theta} = -2\widehat{\theta}^* \widehat{G}_m \widehat{\theta} \\
&= -2 \sum_{j,\ell=1}^m \widehat{\theta}_j \widehat{\theta}_\ell \langle \varphi_j, \varphi_\ell \rangle_{\mathfrak{r},N} = -2\|\widehat{b}_m\|_{\mathfrak{r},N}^2.
\end{aligned}$$

The following theorem provides a risk bound on the adaptive estimator $\widehat{b}_{\widehat{m}}$.

Theorem 2.3 *Assume that $\mathcal{S}_{m'} \subset \mathcal{S}_m$ for every $m, m' \in \mathcal{M}_N$ satisfying $m > m'$. If $\inf_I \mathfrak{f} > 0$ and $b_0 \in \mathbb{L}^4(\mathbb{R}, \mathfrak{f}(x)dx)$, then there exists a constant $\mathfrak{c}_{2.3} > 0$, not depending on N, such that*

$$\mathbb{E}(\|\widehat{b}_{\widehat{m}} - b_0^I\|_{\mathfrak{r},N}^2) \leqslant \mathfrak{c}_{2.3}\left(\min_{m\in\mathcal{M}_N}\left\{\min_{\tau\in\mathcal{S}_m}\|\tau - b_0^I\|_{\mathfrak{f}}^2 + \sigma^2\frac{m}{N}\right\} + \frac{\log(N)^2}{N}\right).$$

Sketch of proof. For a detailed proof of Theorem 2.3, see Comte [2], Theorem 4.1. First of all,

$$\|\widehat{b}_{\widehat{m}} - b_0^I\|_{\mathfrak{r},N}^2 = \underbrace{\|\widehat{b}_{\widehat{m}} - b_0^I\|_{\mathfrak{r},N}^2\mathbf{1}_{\Omega_N}}_{=:A_{\widehat{m}}} + \underbrace{\|\widehat{b}_{\widehat{m}} - b_0^I\|_{\mathfrak{r},N}^2\mathbf{1}_{\Omega_N^c}}_{=:B_{\widehat{m}}}$$

with

$$\Omega_N = \bigcap_{m\in\mathcal{M}_N}\left\{\forall\tau\in\mathcal{S}_m : \frac{1}{\sqrt{2}}\|\tau\|_{\mathfrak{f}} \leqslant \|\tau\|_{\mathfrak{r},N} \leqslant \sqrt{\frac{3}{2}}\|\tau\|_{\mathfrak{f}}\right\}.$$

The sketch of proof is dissected in two steps. Step 1 (resp. Step 2) deals with a suitable bound on $\mathbb{E}(A_{\widehat{m}})$ (resp. $\mathbb{E}(B_{\widehat{m}})$).

Step 1. By the definition of $\widehat{\Pi}_{\widehat{m}}$ (see (2.14)),

$$\|\widehat{b}_{\widehat{m}} - b_0^I\|_{\mathfrak{r},N}^2 = \min_{\tau\in\mathcal{S}_{\widehat{m}}}\|\tau - b_0^I\|_{\mathfrak{r},N}^2 + \|\widehat{b}_{\widehat{m}} - \widehat{\Pi}_{\widehat{m}}(b_0^I)\|_{\mathfrak{r},N}^2.$$

First, $0 \in \mathcal{S}_{\widehat{m}}$ and I is compact, which leads to

$$\min_{\tau\in\mathcal{S}_{\widehat{m}}}\|\tau - b_0^I\|_{\mathfrak{r},N}^2 \leqslant \|b_0^I\|_{\mathfrak{r},N}^2 \leqslant \mathfrak{c}_1 = \sup_{x\in I} b_0(x)^2 < \infty.$$

Now, since $\widehat{m} \in \mathcal{M}_N \subset \{1, \dots, N\}$, and by Equality (2.15),

$$\|\widehat{b}_{\widehat{m}} - \widehat{\Pi}_{\widehat{m}}(b_0^I)\|_{\mathfrak{r},N}^2 \leqslant \sum_{m=1}^N \|\widehat{b}_m - \widehat{\Pi}_m(b_0^I)\|_{\mathfrak{r},N}^2 = \sum_{m=1}^N \widehat{\Delta}_m^*\widehat{G}_m^{-1}\widehat{\Delta}_m.$$

Then,

$$\begin{aligned}\mathbb{E}(\|\widehat{b}_{\widehat{m}} - \widehat{\Pi}_{\widehat{m}}(b_0^I)\|_{\mathfrak{r},N}^2\mathbf{1}_{\Omega_N^c}) &\leqslant \frac{1}{N^2}\sum_{m=1}^N\sum_{j,\ell=1}^m\sum_{i,k=1}^N \mathbb{E}(\varphi_j(\xi_i)\varphi_\ell(\xi_k)\\ &\qquad\times[\widehat{G}_m^{-1}]_{j,\ell}\mathbf{1}_{\Omega_N^c})\mathbb{E}(\varepsilon_i\varepsilon_k)\\ &= \frac{\sigma^2}{N}\sum_{m=1}^N\mathbb{E}\left(\mathbf{1}_{\Omega_N^c}\sum_{j=1}^m[\widehat{G}_m\widehat{G}_m^{-1}]_{j,j}\right) \leqslant \sigma^2 N\mathbb{P}(\Omega_N^c).\end{aligned} \tag{2.17}$$

Moreover, since $\inf_I \mathfrak{f} > 0$ and

$$\mathfrak{L}(m) \leqslant \frac{N}{\log(N)}; \forall m \in \mathcal{M}_N,$$

one can show that there exists a constant $\mathfrak{c}_2 > 0$, not depending on N, such that

$$\mathbb{P}(\Omega_N^c) \leqslant \frac{\mathfrak{c}_2}{N^2} \quad \text{(see Baraud [6])}.$$

Therefore,

$$\mathbb{E}(\|\widehat{b}_{\widehat{m}} - b_0^I\|_{\mathfrak{r},N}^2 \mathbf{1}_{\Omega_N^c}) \leqslant \frac{(\mathfrak{c}_1 + \sigma^2)\mathfrak{c}_2}{N}.$$

Step 2. Consider

$$\rho_N(m) = \mathbb{E}\left(\left(\left[\sup_{\tau \in \mathcal{B}_{m,\widehat{m}}} |\nu_N(\tau)|\right]^2 - p(m, \widehat{m})\right)_+\right); m \in \mathcal{M}_N,$$

where, for every $m' \in \mathcal{M}_N$,

$$\mathcal{B}_{m,m'} = \{\tau \in \mathcal{S}_{m\vee m'} : \|\tau\|_{\mathfrak{f}} = 1\} \quad \text{and} \quad p(m, m') = \frac{\mathfrak{c}_{\text{cal}}}{8} \cdot \frac{m \vee m'}{N}.$$

Consider also

$$\mathfrak{S}_m = \mathcal{S}_1 \cup \cdots \cup \mathcal{S}_m; \forall m \in \{1, \ldots, N\}.$$

First, for every $\tau, \overline{\tau} \in \mathfrak{S}_N$,

$$\gamma_N^{\mathfrak{r}}(\overline{\tau}) - \gamma_N^{\mathfrak{r}}(\tau) = \|\overline{\tau} - b_0^I\|_{\mathfrak{r},N}^2 - \|\tau - b_0^I\|_{\mathfrak{r},N}^2 - 2\nu_N(\overline{\tau} - \tau).$$

Moreover, since

$$\widehat{m} = \arg\min_{m \in \mathcal{M}_N} \{\gamma_N^{\mathfrak{r}}(\widehat{b}_m) + \text{pen}(m)\},$$

for every $m \in \mathcal{M}_N$,

$$\gamma_N^{\mathfrak{r}}(\widehat{b}_{\widehat{m}}) + \text{pen}(\widehat{m}) \leqslant \gamma_N^{\mathfrak{r}}(\widehat{b}_m) + \text{pen}(m).$$

So, for any $m \in \mathcal{M}_N$, since $\mathcal{S}_m + \mathcal{S}_{\widehat{m}} \subset \mathcal{S}_{m \vee \widehat{m}}$,

$$\begin{aligned}\|\widehat{b}_{\widehat{m}} - b_0^I\|_{\mathfrak{r},N}^2 &\leqslant \|\widehat{b}_m - b_0^I\|_{\mathfrak{r},N}^2 + 2\nu_N(\widehat{b}_{\widehat{m}} - \widehat{b}_m) + \text{pen}(m) - \text{pen}(\widehat{m}) \\ &\leqslant \|\widehat{b}_m - b_0^I\|_{\mathfrak{r},N}^2 + \frac{1}{8}\|\widehat{b}_{\widehat{m}} - \widehat{b}_m\|_{\mathfrak{f}}^2 \\ &\quad + 8\left(\left[\sup_{\tau \in \mathcal{B}_{m,\widehat{m}}} |\nu_N(\tau)|\right]^2 - p(m, \widehat{m})\right)_+ \\ &\quad + \text{pen}(m) + 8p(m, \widehat{m}) - \text{pen}(\widehat{m}).\end{aligned}$$

Now, by the definition of Ω_N,

$$\|\tau\|_{\mathfrak{f}}^2 \mathbf{1}_{\Omega_N} \leqslant 2\|\tau\|_{\mathfrak{r},N}^2 \mathbf{1}_{\Omega_N};\ \forall \tau \in \mathfrak{S}_{\max(\mathcal{M}_N)},$$

and since $8p(m, \widehat{m}) \leqslant \text{pen}(m) + \text{pen}(\widehat{m})$,

$$\begin{aligned}\|\widehat{b}_{\widehat{m}} - b_0^I\|_{\mathfrak{r},N}^2 &\leqslant 3\|\widehat{b}_m - b_0^I\|_{\mathfrak{r},N}^2 + 4\text{pen}(m) \\ &\quad + 16\left(\left[\sup_{\tau \in \mathcal{B}_{m,\widehat{m}}} |\nu_N(\tau)|\right]^2 - p(m, \widehat{m})\right)_+ \quad \text{on} \quad \Omega_N.\end{aligned}$$

So,

$$\mathbb{E}(\|\widehat{b}_{\widehat{m}} - b_0^I\|_{\mathfrak{r},N}^2 \mathbf{1}_{\Omega_N}) \leqslant \min_{m \in \mathcal{M}_N} \{3\mathbb{E}(\|\widehat{b}_m - b_0^I\|_{\mathfrak{r},N}^2) + 4\text{pen}(m) + 16\rho_N(m)\}.$$

Thanks to Talagrand's inequality (see Klein and Rio [7]), one can show that there exists a constant $\mathfrak{c}_3 > 0$, not depending on N, such that

$$\sup_{m \in \mathcal{M}_N} \rho_N(m) \leqslant \mathfrak{c}_3 \frac{\log(N)^2}{N} \quad \text{(see Comte [13], Theorem 4.1).}$$

Therefore,

$$\begin{aligned}\mathbb{E}(\|\widehat{b}_{\widehat{m}} - b_0^I\|_{\mathfrak{r},N}^2) &\leqslant \min_{m \in \mathcal{M}_N} \left\{3\mathbb{E}(\|\widehat{b}_m - b_0^I\|_{\mathfrak{r},N}^2) + 4\mathfrak{c}_{\text{cal}}\sigma^2 \frac{m}{N}\right\} \\ &\quad + \underbrace{\frac{(\mathfrak{c}_1 + \sigma^2)\mathfrak{c}_2}{N} + 16\mathfrak{c}_3 \frac{\log(N)^2}{N}}_{\lesssim \frac{\log(N)^2}{N}},\end{aligned}$$

and the conclusion comes from Proposition 2.3.

Let us conclude this section with some remarks about Theorem 2.3 and Chap. 3:

- If I is not compact, $\widehat{m}$ needs to be selected in a random subset of $\{1, \dots, N\}$ depending on the lowest eigenvalue of $\widehat{G}_m$ ($m \in \{1, \dots, N\}$), and then there are additional steps in the proof of Theorem 2.3 (see Comte and Genon-Catalot [8], Theorem 2). In Chap. 3, for the projection least squares estimator of b_0 computed from the copies $X^1, \dots, X^N$ of the diffusion process X, I may be not compact.
- To establish risk bounds on the projection least squares estimator of b_0 computed from $X^1, \dots, X^N$ is of course more difficult than in the nonparametric regression framework, especially because the noise component in Eq. (1.1) depends on X, which doesn't allow to directly prove equalities as (2.16) and (2.17).

References

1. Lacour, C., Massart, P. and Rivoirard, V. (2017). Estimator Selection: a New Method with Applications to Kernel Density Estimation. *Sankhya A* **79**(2), 298–335.
2. Comte, F. (2014). *Estimation non-paramétrique*. Spartacus IDH, France.
3. Halconruy, H. and Marie, N. (2020). Kernel Selection in Nonparametric Regression. *Mathematical Methods of Statistics* **29**(1), 31–55.
4. Houdré, C. and Reynaud-Bouret, P. (2003). Exponential Inequalities, with Constants, for U-Statistics of Order Two. *Stochastic Inequalities and Applications*, Progress in Probability **56**, 55–69, Birkhäuser, Basel.
5. DeVore, R.A. and Lorentz, G.G. (1993). *Constructive Approximation*. Springer, Berlin-Heidelberg.
6. Baraud, Y. (2002). Model Selection for Regression on a Random Design. *ESAIM:PS* **6**, 127–146.
7. Klein, T. and Rio, E. (2005). Concentration Around the Mean for Maxima of Empirical Processes. *The Annals of Probability* **33**(3), 1060–1077.
8. Comte, F. and Genon-Catalot, V. (2020). Regression Function Estimation as a Partly Inverse Problem. *The Annals of the Institute of Statistical Mathematics* **72**(4), 1023–1054.

Chapter 3
The Projection Least Squares Estimator of the Drift Function

This chapter deals with risk bounds on the following truncated projection least squares estimator of b_0:

$$\widetilde{b}_m(x) = \widehat{b}_m(x)\mathbf{1}_{\Lambda_m}; \, x \in I,$$

where

$$\Lambda_m = \left\{\mathfrak{L}(m)(\|\widehat{\Psi}_m^{-1}\|_{\text{op}} \vee 1) \leqslant \mathfrak{c}_\Lambda \frac{NT}{\log(NT)}\right\},$$

the deterministic constant $\mathfrak{c}_\Lambda > 0$ doesn't depend on m and N but is related to the dependence structure of $X^1, \dots, X^N$, and

$$\mathfrak{L}(m) = 1 \vee \left(\sup_{x \in I} \sum_{j=1}^{m} \varphi_j(x)^2\right).$$

First of all, Sects. 3.1 and 3.2 are brief reminders on Itô's integral and on symmetric (random) matrices respectively. In a more general framework than that of copies of a diffusion process, Sects. 3.3 and 3.4 answer the following two questions:

- Are the empirical norm and its theoretical counterpart equivalent on $\mathcal{S}_m$?
- Is the truncated estimator consistent?

In Sect. 3.5, a risk bound on $\widetilde{b}_m$ is established for fixed m. Finally, Sect. 3.6 deals with a model selection method and a risk bound on the associated adaptive estimator. Throughout this chapter, σ satisfies (1.3). Then, for every $t \in (0, T]$, the probability distribution of X_t has a density f_t with respect to Lebesgue measure such that, for every $x \in \mathbb{R}$,

$$f_t(x) \leqslant \mathfrak{c}_{0.5,T} t^{-\frac{1}{2}} \exp\left(-\mathfrak{m}_{0.5,T} \frac{(x - x_0)^2}{t}\right) \tag{3.1}$$

N. Marie, *From Nonparametric Regression to Statistical Inference for Non-Ergodic Diffusion Processes*, Frontiers in Probability and the Statistical Sciences,
https://doi.org/10.1007/978-3-031-95638-6_3

where $\mathfrak{c}_{0.5,T}$ and $\mathfrak{m}_{0.5,T}$ are positive constants depending on T but not on t and x (see Menozzi et al. [1], Theorem 1.2). So, $t \mapsto f_t(x)$ ($x \in \mathbb{R}$) belongs to $\mathbb{L}^1([0, T])$, which legitimates to consider the density function f defined by

$$f(x) = \frac{1}{T}\int_0^T f_s(x)ds;\ \forall x \in \mathbb{R}.$$

Still by Inequality (3.1):

- Since b_0' is bounded (and then b_0 has linear growth),

$$|b_0|^\kappa \in \mathbb{L}^2(\mathbb{R}, f(x)dx);\ \forall \kappa \in \mathbb{R}_+.$$

- f is bounded. So, $\mathcal{S}_m \subset \mathbb{L}^2(\mathbb{R}, f(x)dx)$ because the φ_j's belong to $\mathbb{L}^2(\mathbb{R}, dx)$.

Remarks:

- For every $t \in (0, T]$, let μ_t be the probability distribution of X_t. By the Fubini-Tonelli theorem, for every piecewise continuous functions $\varphi, \psi : \mathbb{R} \to \mathbb{R}$,

$$\mathbb{E}(\langle \varphi, \psi \rangle_N) = \int_{-\infty}^{\infty} \varphi(x)\psi(x)\mu(dx) \quad \text{with} \quad \mu(dx) = \frac{1}{T}\int_0^T \mu_s(dx)ds,$$

  and then one could think at first sight that μ_t doesn't need to be absolutely continuous with respect to Lebesgue's measure for any $t \in (0, T]$. However, since b_0 may be unbounded, a sharp bound on μ_t as (3.1) is required to show that

$$|b_0|^\kappa \in \mathbb{L}^2(\mathbb{R}, \mu);\ \forall \kappa \in \mathbb{R}_+.$$

- Let $t \mapsto \theta_t(x_0)$ be the solution of the differential equation

$$x_t = x_0 + \int_0^t b_0(x_s)ds;\ t \in [0, T].$$

  Precisely, Menozzi et al. [1], Theorem 1.2 says that there exist two constants $\mathfrak{c}_1, \mathfrak{c}_2 > 0$ such that, for every $t \in (0, T]$ and $x \in \mathbb{R}$,

$$f_t(x) \leqslant \mathfrak{c}_1 t^{-\frac{1}{2}} \exp\left(-\mathfrak{c}_2 \frac{(x - \theta_t(x_0))^2}{t}\right). \tag{3.2}$$

  For every $t \in (0, T]$ and $x \in \mathbb{R}$,

$$|x - x_0| \leqslant |x - \theta_t(x_0)| + tS_T(x_0) \\ \quad\quad \text{with} \quad S_T(x_0) = \sup_{s\in[0,T]} |b_0(\theta_s(x_0))|,$$

leading to

$$-\frac{1}{2}(x-x_0)^2+t^2S_T(x_0)^2 \geqslant -(x-\theta_t(x_0))^2.$$

Thus, Inequality (3.1) is a straightforward consequence of Inequality (3.2).

3.1 A Brief Reminder on Itô's Integral

Throughout this section, $(\Omega, \mathcal{F}, \mathbb{P})$ is a probability space, $\mathbb{F} = (\mathcal{F}_t)_{t\in[0,T]}$ is a filtration and $W = (W_t)_{t\in[0,T]}$ is a $\mathbb{F}$-adapted Brownian motion. The compensator $\langle W\rangle = (\langle W\rangle_t)_{t\in[0,T]}$ of W, which coincides with its quadratic variation because W is a continuous martingale, satisfies $\langle W\rangle_t = t$ for every $t \in [0, T]$. In order to define the (Itô) stochastic integral, let us introduce the two following function spaces:

$$\mathcal{E} = \left\{ \sum_{i=0}^{n-1} \xi_i \mathbf{1}_{[t_i,t_{i+1}[};\, 0 = t_0 < t_1 < \cdots < t_n = T \text{ with } n \in \mathbb{N}^*, \text{ and} \right.$$
$$\left. \xi_0, \ldots, \xi_{n-1} \in \mathbb{L}^2(\Omega) \text{ s.t. } \xi_i \text{ is } \mathcal{F}_{t_i}\text{-measurable} \right\}$$

and

$$\mathbb{H}^2 = \{H \in \mathbb{L}^2(\Omega \times [0, T]) : H \text{ is } \mathbb{F}\text{-adapted}\} = \overline{\mathcal{E}}^{N_2},$$

where N_2 is the usual norm on $\mathbb{L}^2(\Omega \times [0, T])$. The stochastic integral on $[0, T]$ of $H = \sum_{i=0}^{n-1} \xi_i \mathbf{1}_{[t_i,t_{i+1}[} \in \mathcal{E}$ with respect to W is defined by

$$\int_0^T H_s dW_s = \sum_{i=0}^{n-1} \xi_i (W_{t_{i+1}} - W_{t_i}).$$

It satisfies the two following properties which are crucial in the sequel: for every $H \in \mathcal{E}$,

(A) $$\int_0^{\cdot} H_s dW_s = \left(\int_0^T H_s \mathbf{1}_{[0,t]}(s) dW_s \right)_{t\in[0,T]} \text{ is a } \mathbb{F}\text{-martingale,}$$

(B) $$\text{and } \mathbb{E}\left[\left(\int_0^T H_s dW_s \right)^2 \right] = \int_0^T \mathbb{E}(H_s^2) ds \text{ (isometry property).}$$

Thanks to the density result $\overline{\mathcal{E}}^{N_2} = \mathbb{H}^2$, there is a unique extension of the stochastic integral with respect to W from $\mathcal{E}$ to $\mathbb{H}^2$ still satisfying the martingale property (A) and the isometry property (B). Moreover, for every $H \in \mathbb{H}^2$,

(C) $$\left\langle \int_0^{\cdot} H_s dW_s \right\rangle_T = \int_0^T H_s^2 d\langle W\rangle_s = \int_0^T H_s^2 ds.$$

Finally, the following consequence of the Burkholder-Davis-Gundy inequality is also useful in the sequel:

(D) For every $p \geqslant 1$, there exists a constant $\mathfrak{b}_p > 0$ such that, for every $H \in \mathbb{H}^2$,

$$\mathbb{E}\left[\sup_{t\in[0,T]}\left|\int_0^t H_s dW_s\right|^p\right] \leqslant \mathfrak{b}_p \mathbb{E}\left[\left(\int_0^T H_s^2 ds\right)^{\frac{p}{2}}\right].$$

For details on Itô's integral, the reader may refer to Revuz and Yor [2], Chap. IV.

3.2 A Brief Reminder on Symmetric (Random) Matrices

First of all, let us define the Loewner order and recall some basic properties of (positive semidefinite) symmetric matrices.

Definition 3.1 For every $d \times d$ symmetric matrices A and B, $A \preccurlyeq B$ (resp. $A \prec B$) if and only if,

$$x^* A x \leqslant x^* B x \quad (\text{resp. } x^* A x < x^* B x)); \forall x \in \partial B_{2,d}(0, 1).$$

Proposition 3.1 *Let A be a $d \times d$ symmetric matrix.*

1. *For every $d \times d$ symmetric matrix B, and every $d \times d$ invertible matrix M,*

$$A \preccurlyeq B \Longleftrightarrow MAM^* \preccurlyeq MBM^*.$$

2. *If $A \succcurlyeq 0$, then there exists a unique $d \times d$ symmetric matrix $A^{1/2}$, called the square root of A, such that*

$$A^{\frac{1}{2}} \succcurlyeq 0 \quad \textit{and} \quad A^{\frac{1}{2}} A^{\frac{1}{2}} = A.$$

3. *Consider $\lambda_{\min}, \lambda_{\max} \in \mathbb{R}$ such that $\lambda_{\min} < \lambda_{\max}$. All the eigenvalues of A belong to $[\lambda_{\min}, \lambda_{\max}]$ if and only if $\lambda_{\min} I \preccurlyeq A \preccurlyeq \lambda_{\max} I$.*
4. *The lowest (resp. largest) eigenvalue of the matrix A is denoted by $\lambda_{\min}(A)$ (resp. $\lambda_{\max}(A)$) and satisfies*

$$\lambda_{\min}(A) = \inf_{x:\|x\|_{2,d}=1} x^* A x \quad (\textit{resp. } \lambda_{\max}(A) = \sup_{x:\|x\|_{2,d}=1} x^* A x).$$

Moreover,

$$\|A\|_{\text{op}} = \max\{-\lambda_{\min}(A), \lambda_{\max}(A)\} = \sup_{x:\|x\|_{2,d}=1} |x^* A x|.$$

5. *If $A \succcurlyeq 0$ and A is invertible, then*

$$\lambda_{\max}(A^{-1}) = \frac{1}{\lambda_{\min}(A)} \quad \textit{and} \quad \lambda_{\min}(A^{-1}) = \frac{1}{\lambda_{\max}(A)}.$$

6. *If* $A \succcurlyeq 0$, *then*

$$\mathrm{trace}(AB) \leqslant \|B\|_{\mathrm{op}}\mathrm{trace}(A);\ \forall B \in \mathcal{M}_d(\mathbb{R}).$$

Convention. In the sequel, if a $d \times d$ symmetric matrix A such that $A \succcurlyeq 0$ is not invertible, then $\|A^{-1}\|_{\mathrm{op}} = \infty$. This convention makes sense thanks to Proposition 3.1.(5).

Now, let us state three concentration inequalities for symmetric random matrices.

Proposition 3.2 (Matrix Chernoff's inequality) *Let A be a $d \times d$ symmetric random matrix such that*

$$A \succcurlyeq 0 \quad \textit{and} \quad \mathbb{P}(\lambda_{\max}(A) \leqslant R) = 1,$$

where $R > 0$ is a deterministic constant. Consider $S = A_1 + \cdots + A_n$, where $n \in \mathbb{N}^*$ and $A_1, \ldots, A_n$ are independent copies of A. Then,

$$\mathbb{P}(\lambda_{\min}(S) \leqslant (1-\delta)\lambda_{\min}(\mathbb{E}(S))) \leqslant d\left(\frac{e^{-\delta}}{(1-\delta)^{1-\delta}}\right)^{\frac{\lambda_{min}(\mathbb{E}(S))}{R}};\ \forall \delta \in [0,1],$$

and

$$\mathbb{P}(\lambda_{\max}(S) \geqslant (1+\delta)\lambda_{\max}(\mathbb{E}(S))) \leqslant d\left(\frac{e^{\delta}}{(1+\delta)^{1+\delta}}\right)^{\frac{\lambda_{\max}(\mathbb{E}(S))}{R}};\ \forall \delta \geqslant 0.$$

See Tropp [3], Theorem 1.1 for a proof.

Proposition 3.3 (Matrix Bernstein's inequality) *Let A be a $d \times d$ symmetric random matrix such that*

$$\mathbb{E}(A) = 0 \quad \textit{and} \quad \mathbb{P}(\|A\|_{\mathrm{op}} \leqslant R) = 1,$$

where $R > 0$ is a deterministic constant. Consider $S = A_1 + \cdots + A_n$, where $n \in \mathbb{N}^*$ and $A_1, \ldots, A_n$ are independent copies of A. Then, for every $\delta \geqslant 0$,

$$\mathbb{P}(\|S\|_{\mathrm{op}} \geqslant \delta) \leqslant d\ \exp\left(-\frac{\delta^2/2}{\sigma^2 + R\delta/3}\right) \quad \textit{with} \quad \sigma^2 = N\|\mathbb{E}(A^2)\|_{\mathrm{op}}.$$

See Tropp [3], Theorem 1.4 for a proof.

Proposition 3.4 (Matrix Azuma's inequality) *Let A be a centered $d \times d$ symmetric random matrix such that*

$$\mathbb{P}(A^2 \preccurlyeq M^2) = 1,$$

where M is a $d \times d$ symmetric (deterministic) matrix. Consider $S = A_1 + \cdots + A_n$, where $n \in \mathbb{N}^*$ and $A_1, \ldots, A_n$ are copies of A such that

$$\mathbb{E}(A_i | A_{i-1}, \ldots, A_1) = 0; \forall i \in \{2, \ldots, n\}.$$

Then, for every $\delta \geqslant 0$,

$$\mathbb{P}(\lambda_{\max}(S) \geqslant \delta) \leqslant d e^{-\frac{\delta^2}{8\sigma^2}} \quad with \quad \sigma^2 = n\|M^2\|_{\text{op}}.$$

See Tropp [3], Theorem 7.1 for a proof.

3.3 Are the Empirical Norm and Its Theoretical Counterpart Equivalent on the Projection Space?

Let $\Gamma = (\Gamma_t)_{t \in [0,T]}$ be a piecewise continuous stochastic process, and assume that for every $t \in (0, T]$, the probability distribution of Γ_t has a density γ_t with respect to Lebesgue's measure such that $t \mapsto \gamma_t(x)$ ($x \in \mathbb{R}$) belongs to $\mathbb{L}^1([0, T], dt)$. This legitimates to consider the density function γ defined by

$$\gamma(x) = \frac{1}{T}\int_0^T \gamma_s(x)ds; \forall x \in \mathbb{R}.$$ Moreover, assume that $\mathcal{S}_m \subset \mathbb{L}^2(\mathbb{R}, \gamma(x)dx)$.

Consider also N copies $\Gamma^1, \ldots, \Gamma^N$ of Γ, the random matrix $\widehat{\Psi}_{m,\Gamma} = (\langle \varphi_j, \varphi_\ell \rangle_{N,\Gamma})_{j,\ell}$ with

$$\langle \varphi, \psi \rangle_{N,\Gamma} := \frac{1}{NT}\sum_{i=1}^{N}\int_0^T \varphi(\Gamma_s^i)\psi(\Gamma_s^i)ds,$$

and the matrix $\Psi_{m,\gamma} = (\langle \varphi_j, \varphi_\ell \rangle_\gamma)_{j,\ell}$ with

$$\langle \varphi, \psi \rangle_\gamma := \int_{-\infty}^{\infty} \varphi(x)\psi(x)\gamma(x)dx.$$

In order to establish a risk bound on the projection least squares estimator of b_0, $\|.\|_{N,\Gamma}$ and $\|.\|_\gamma$ need to be almost surely equivalent on $\mathcal{S}_m$ when $N \to \infty$. To that purpose, let us introduce the event

$$\begin{aligned}\Omega_{m,\Gamma} &= \left\{\forall \tau \in \mathcal{S}_m : \frac{1}{\sqrt{2}}\|\tau\|_\gamma \leqslant \|\tau\|_{N,\Gamma} \leqslant \sqrt{\frac{3}{2}}\|\tau\|_\gamma\right\} \\ &= \left\{\sup_{\tau \in \mathcal{S}_m : \|\tau\|_\gamma = 1} |\|\tau\|_{N,\Gamma}^2 - \|\tau\|_\gamma^2| \leqslant \frac{1}{2}\right\}.\end{aligned}$$

This section deals with a suitable bound on $\mathbb{P}(\Omega_{m,\Gamma}^c)$ when

$$\Gamma^1, \dots, \Gamma^N \text{ are independent (see Sect. 3.1.1),}$$

and then when

$$\Gamma^1, \dots, \Gamma^N \text{ have a } \textit{sparse} \text{ dependence structure (see Sect. 3.3.2).}$$

Two technical results must be established first.

Lemma 3.1 *If the (positive semidefinite symmetric) matrix* $\Psi_{m,\Gamma}$ *is invertible, then*

$$\Omega_{m,\Gamma} = \left\{ \|\widehat{G}_{m,\Gamma} - I\|_{\mathrm{op}} \leqslant \frac{1}{2} \right\} \quad \textit{with} \quad \widehat{G}_{m,\Gamma} = \Psi_{m,\gamma}^{-\frac{1}{2}} \widehat{\Psi}_{m,\Gamma} \Psi_{m,\gamma}^{-\frac{1}{2}}.$$

Proof For any $\tau = \sum_{j=1}^{m} \theta_j \varphi_j \in \mathcal{S}_m$,

$$\|\tau\|_{N,\Gamma}^2 = \frac{1}{NT} \sum_{i=1}^{N} \int_0^T \left(\sum_{j=1}^{m} \theta_j \varphi_j(\Gamma_s^i) \right)^2 ds = \sum_{j,\ell=1}^{m} \theta_j \theta_\ell [\widehat{\Psi}_{m,\Gamma}]_{j,\ell} = \theta^* \widehat{\Psi}_{m,\Gamma} \theta$$

and, in the same way, $\|\tau\|_\gamma^2 = \theta^* \Psi_{m,\gamma} \theta$. Moreover, since the symmetric matrix $\Psi_{m,\gamma}$ is positive semidefinite,

$$\|\tau\|_\gamma^2 = \theta^* (\Psi_{m,\gamma}^{\frac{1}{2}})^* \Psi_{m,\gamma}^{\frac{1}{2}} \theta = \|\Psi_{m,\gamma}^{\frac{1}{2}} \theta\|_{2,m}^2.$$

Therefore, by Proposition 3.1.(4), and since $\Psi_{m,\gamma}$ is invertible,

$$\begin{aligned}
\sup_{\tau \in \mathcal{S}_m : \|\tau\|_\gamma = 1} \left| \|\tau\|_{N,\Gamma}^2 - \|\tau\|_\gamma^2 \right| &= \sup_{\theta : \|\Psi_{m,\gamma}^{1/2} \theta\|_{2,m} = 1} |\theta^* (\widehat{\Psi}_{m,\Gamma} - \Psi_{m,\gamma}) \theta| \\
&= \sup_{\theta : \|\theta\|_{2,m} = 1} |\theta^* \Psi_{m,\gamma}^{-\frac{1}{2}} (\widehat{\Psi}_{m,\Gamma} - \Psi_{m,\gamma}) \Psi_{m,\gamma}^{-\frac{1}{2}} \theta| \\
&= \sup_{\theta : \|\theta\|_{2,m} = 1} |\theta^* (\Psi_{m,\gamma}^{-\frac{1}{2}} \widehat{\Psi}_{m,\Gamma} \Psi_{m,\gamma}^{-\frac{1}{2}} - I) \theta| = \|\widehat{G}_{m,\Gamma} - I\|_{\mathrm{op}}.
\end{aligned}$$

□

Now, by assuming that $\Psi_{m,\gamma}$ is invertible, consider the functions $\overline{\varphi}_1, \dots, \overline{\varphi}_m$ defined by

$$\overline{\varphi}_j = \sum_{\ell=1}^{m} [\Psi_{m,\gamma}^{-\frac{1}{2}}]_{j,\ell} \varphi_\ell ; \forall j \in \{1, \dots, m\}.$$

Moreover, note that

$$\widehat{G}_{m,\Gamma} = \sum_{i=1}^{N} G_{m,\gamma}(\Gamma^i)$$

where, for every piecewise continuous function $h : [0, T] \to \mathbb{R}$,

$$G_{m,\gamma}(h) = \frac{1}{N}\Psi_{m,\gamma}^{-\frac{1}{2}}\Psi_m(h)\Psi_{m,\gamma}^{-\frac{1}{2}}$$

and

$$\Psi_m(h) = \left(\frac{1}{T}\int_0^T \varphi_j(h(s))\varphi_\ell(h(s))ds\right)_{j,\ell\in\{1,\dots,m\}}.$$

Lemma 3.2 *If the (positive semidefinite symmetric) matrix $\Psi_{m,\gamma}$ is invertible, then $(\overline{\varphi}_1, \dots, \overline{\varphi}_m)$ is an orthonormal basis of $\mathcal{S}_m$ in $\mathbb{L}^2(\mathbb{R}, \gamma(x)dx)$.*

Proof On the one hand, since $\mathbb{E}(\Psi_m(\Gamma)) = \Psi_{m,\gamma}$,

$$\mathbb{E}(G_{m,\gamma}(\Gamma)) = \frac{1}{N}\Psi_{m,\gamma}^{-\frac{1}{2}}\mathbb{E}(\Psi_m(\Gamma))\Psi_{m,\gamma}^{-\frac{1}{2}} = \frac{1}{N}\Psi_{m,\gamma}^{-\frac{1}{2}}\Psi_{m,\gamma}\Psi_{m,\gamma}^{-\frac{1}{2}} = \frac{1}{N}I.$$

On the other hand, for every piecewise continuous function $h : [0, T] \to \mathbb{R}$,

$$\begin{aligned} G_{m,\gamma}(h) &= \frac{1}{NT}\int_0^T \left(\sum_{\ell,\ell'=1}^m [\Psi_{m,\gamma}^{-\frac{1}{2}}]_{j,\ell}[\Psi_{m,\gamma}^{-\frac{1}{2}}]_{j',\ell'}\varphi_\ell(h(s))\varphi_{\ell'}(h(s))\right)_{j,j'} ds \\ &= \frac{1}{NT}\int_0^T (\overline{\varphi}_j(h(s))\overline{\varphi}_{j'}(h(s)))_{j,j'}ds. \end{aligned}$$

Then,

$$\begin{aligned} \mathbb{E}(G_{m,\gamma}(\Gamma)) &= \frac{1}{NT}\int_0^T \left(\int_I \overline{\varphi}_j(x)\overline{\varphi}_{j'}(x)\gamma_s(x)dx\right)_{j,j'} ds \\ &= \frac{1}{N}\left(\int_I \overline{\varphi}_j(x)\overline{\varphi}_{j'}(x)\gamma(x)dx\right)_{j,j'} = \frac{1}{N}(\langle\overline{\varphi}_j, \overline{\varphi}_{j'}\rangle_\gamma)_{j,j'}. \end{aligned}$$

Therefore, $(\overline{\varphi}_1, \dots, \overline{\varphi}_m)$ is an orthonormal basis of $\mathcal{S}_m$ in $\mathbb{L}^2(\mathbb{R}, \gamma(x)dx)$. □

3.3.1 Case 1: Independent Observations

In this section, the copies $\Gamma^1, \dots, \Gamma^N$ of Γ are independent,

$$\mathfrak{c}_\Lambda = \frac{1-\log(2)}{(1+p)T} \quad \text{with} \quad p \in \mathbb{N}^*,$$

and the dimension m of $\mathcal{S}_m$ fulfills the following assumption.

Assumption 3.1 The dimension m of $\mathcal{S}_m$ satisfies

$$\mathfrak{L}(m)^{-1}(\lambda_{\min}(\Psi_{m,\gamma}) \wedge 1) \geqslant \frac{1}{\overline{\mathfrak{c}}_\Lambda}\cdot\frac{\log(NT)}{NT} \quad \text{with} \quad \overline{\mathfrak{c}}_\Lambda = \frac{\mathfrak{c}_\Lambda}{2}.$$

Let us make some remarks about Assumption 3.1:

- Assumption 3.1 has been introduced in the nonparametric regression framework by A. Cohen, M.A. Davenport and D. Leviatan (see [4]) in order to improve the stability of the projection least squares estimator of b_0.
- Under Assumption 3.1,

$$\lambda_{\min}(\Psi_{m,\gamma}) \geqslant \lambda_{\min}(\Psi_{m,\gamma}) \wedge 1 \geqslant \frac{\mathfrak{L}(m)}{\bar{\mathfrak{c}}_\Lambda} \cdot \frac{\log(NT)}{NT} > 0.$$

Then, $\Psi_{m,\gamma}$ is invertible, and since $\|\Psi_{m,\gamma}^{-1}\|_{\mathrm{op}} = \lambda_{\min}(\Psi_{m,\gamma})^{-1}$ by Proposition 3.1.(5), Assumption 3.1 may be reformulated a more convenient way:

$$\mathfrak{L}(m)(\|\Psi_{m,\gamma}^{-1}\|_{\mathrm{op}} \vee 1) \leqslant \bar{\mathfrak{c}}_\Lambda \frac{NT}{\log(NT)}.$$

- Under Assumption 3.1, the set of *authorized* m is smaller than $\{1,\dots,N_T\}$, but how much? Let us answer this question when $I = [\mathtt{l},\mathtt{r}]$ with $\mathtt{l},\mathtt{r} \in \mathbb{R}$ satisfying $\mathtt{l} < \mathtt{r}$. First, assume that there exists a constant $\underline{\mathfrak{m}} > 0$ such that $\gamma(\cdot) \geqslant \underline{\mathfrak{m}}$ on I, which is true for $\gamma = f$ thanks to Menozzi et al. [1], Theorem 1.2. Then,

$$\begin{aligned}\|\Psi_{m,\gamma}^{-1}\|_{\mathrm{op}} = \frac{1}{\lambda_{\min}(\Psi_{m,\gamma})} &= \left(\inf_{\theta:\|\theta\|_{2,m}=1} \sum_{j,\ell=1}^{m} \theta_j\theta_\ell[\Psi_{m,\gamma}]_{j,\ell}\right)^{-1} \\ &= \left[\inf_{\theta:\|\theta\|_{2,m}=1} \int_{\mathtt{l}}^{\mathtt{r}} \left(\sum_{j=1}^{m} \theta_j\varphi_j(x)\right)^2 \gamma(x)dx\right]^{-1} \leqslant \frac{1}{\underline{\mathfrak{m}}}.\end{aligned}$$

Moreover, assume that there exists a constant $\mathfrak{c}_\varphi > 0$, not depending on m and N, such that

$$\mathfrak{L}(m) \leqslant \mathfrak{c}_\varphi^2 m. \tag{3.3}$$

Then, Assumption 3.1 is fulfilled by any

$$m \leqslant \frac{\bar{\mathfrak{c}}_\Lambda}{\mathfrak{c}_\varphi^2(\underline{\mathfrak{m}}^{-1} \vee 1)} \cdot \frac{NT}{\log(NT)}.$$

For instance, the $[\mathtt{l},\mathtt{r}]$-supported trigonometric basis satisfies (3.3) because $\cos(\cdot)^2 + \sin(\cdot)^2 = 1$.

The following proposition provides a suitable bound on $\mathbb{P}(\Omega_{m,\Gamma}^c)$ under Assumption 3.1.

Proposition 3.5 *Under Assumption 3.1,*

$$\mathbb{P}(\Omega_{m,\Gamma}^c) \leqslant \frac{\mathfrak{c}_{3.5}}{N^p} \quad \textit{with} \quad \mathfrak{c}_{3.5} = \frac{2}{T^{p+1}}.$$

Proof In order to control $\mathbb{P}(\Omega_{m,\Gamma}^c)$ via the matrix Chernoff inequality (see Proposition 3.2), let us establish that for any piecewise continuous function $h : [0, T] \to \mathbb{R}$,

$$G_{m,\gamma}(h) \succcurlyeq 0 \quad \text{and} \quad \lambda_{\max}(G_{m,\gamma}(h)) \leqslant R = \frac{1}{N}\mathfrak{L}(m)\|\Psi_{m,\gamma}^{-1}\|_{\text{op}}.$$

On the one hand, as established in the proof of Lemma 3.2,

$$G_{m,\gamma}(h) = \frac{1}{NT}\int_0^T (\overline{\varphi}_j(h(s))\overline{\varphi}_\ell(h(s)))_{j,\ell}ds$$

and then, for every $x \in \mathbb{R}^m$,

$$x^*G_{m,\gamma}(h)x = \sum_{j,\ell=1}^m G_{m,\gamma}(h)_{j,\ell}x_jx_\ell = \frac{1}{NT}\int_0^T\left(\sum_{j=1}^m \overline{\varphi}_j(h(s))x_j\right)^2 ds \geqslant 0.$$

On the other hand, by Jensen's and Cauchy-Schwarz's inequalities,

$$\begin{aligned}
\|\Psi_m(h)\|_{\text{op}}^2 &= \sup_{x:\|x\|_{2,m}=1}\sum_{\ell=1}^m\left(\frac{1}{T}\int_0^T\left(\sum_{j=1}^m \varphi_j(h(s))\varphi_\ell(h(s))x_j\right)ds\right)^2 \\
&\leqslant \frac{1}{T}\sup_{x:\|x\|_{2,m}=1}\sum_{\ell=1}^m\int_0^T \varphi_\ell(h(s))^2\left(\sum_{j=1}^m \varphi_j(h(s))x_j\right)^2 ds \\
&\leqslant \frac{\mathfrak{L}(m)}{T}\sup_{x:\|x\|_{2,m}=1}\int_0^T\left(\sum_{j=1}^m \varphi_j(h(s))x_j\right)^2 ds \leqslant \mathfrak{L}(m)^2 \qquad (3.4)
\end{aligned}$$

and then, since $G_{m,\gamma}(h)$ is a positive semidefinite symmetric matrix, by Proposition 3.1.(3,4),

$$\begin{aligned}
\lambda_{\max}(G_{m,\gamma}(h)) &= \|G_{m,\gamma}(h)\|_{\text{op}} \\
&\leqslant \frac{1}{N}\|\Psi_m(h)\|_{\text{op}}\|\Psi_{m,\gamma}^{-1}\|_{\text{op}} \leqslant \frac{1}{N}\mathfrak{L}(m)\|\Psi_{m,\gamma}^{-1}\|_{\text{op}}. \qquad (3.5)
\end{aligned}$$

So, by the matrix Chernoff inequality, for every $\delta \in [0, 1]$,

$$\mathbb{P}(\lambda_{\min}(\widehat{G}_{m,\Gamma}) \leqslant (1-\delta)\lambda_{\min}(\mathbb{E}(\widehat{G}_{m,\Gamma}))) \leqslant m\left(\frac{e^{-\delta}}{(1-\delta)^{1-\delta}}\right)^{\frac{\lambda_{\min}(\mathbb{E}(\widehat{G}_{m,\Gamma}))}{R}},$$

and for every $\delta \geqslant 0$,

$$\mathbb{P}(\lambda_{\max}(\widehat{G}_{m,\Gamma}) \geqslant (1+\delta)\lambda_{\max}(\mathbb{E}(\widehat{G}_{m,\Gamma}))) \leqslant m\left(\frac{e^{\delta}}{(1+\delta)^{1+\delta}}\right)^{\frac{\lambda_{\max}(\mathbb{E}(\widehat{G}_{m,\Gamma}))}{R}}.$$

Moreover, since

$$\mathbb{E}(\widehat{G}_{m,\Gamma}) = N\mathbb{E}(G_{m,\gamma}(\Gamma)) = I$$

as established in the proof of Lemma 3.2, the smallest and the largest eigenvalues of $\mathbb{E}(\widehat{G}_{m,\Gamma})$ are both equal to 1. Thus, by Proposition 3.1.(4), and since

$$\mathrm{spec}(\widehat{G}_{m,\Gamma} - I) = \{\lambda - 1; \lambda \in \mathrm{spec}(\widehat{G}_{m,\Gamma})\},$$

for every $\delta \in (0, 1)$,

$$\begin{aligned}\mathbb{P}(\|\widehat{G}_{m,\Gamma} - I\|_{\mathrm{op}} > \delta) &\leqslant \mathbb{P}(\lambda_{\min}(\widehat{G}_{m,\Gamma}) \leqslant 1-\delta) + \mathbb{P}(\lambda_{\max}(\widehat{G}_{m,\Gamma}) \geqslant 1+\delta)\\ &\leqslant m\left(\frac{e^{-\delta}}{(1-\delta)^{1-\delta}}\right)^{\frac{1}{R}}\\ &\quad + m\left(\frac{e^{\delta}}{(1+\delta)^{1+\delta}}\right)^{\frac{1}{R}} \leqslant 2m\left(\frac{e^{-\delta}}{(1-\delta)^{1-\delta}}\right)^{\frac{1}{R}}.\end{aligned}$$

Finally, for $\delta = 1/2$ and by Assumption 3.1,

$$\begin{aligned}2m\left(\frac{e^{-\delta}}{(1-\delta)^{1-\delta}}\right)^{\frac{1}{R}} &= 2m\exp\left(-\frac{1}{2R}(1-\log(2))\right)\\ &\leqslant 2m\exp\left(-\frac{\log(NT)}{\mathfrak{c}_{\Lambda}T}(1-\log(2))\right) = \frac{2m}{(NT)^{p+1}} \leqslant \frac{\mathfrak{c}_1}{N^p}\end{aligned}$$

with $\mathfrak{c}_1 = 2T^{-(p+1)}$. Thanks to Lemma 3.1, this concludes the proof. □

3.3.2 *Case 2: Sparse Dependent Observations*

In this section, $\Gamma^1, \dots, \Gamma^N$ have a *sparse* dependence structure in the following sense.

Assumption 3.2 Consider the set

$$\mathcal{I}_{N,\Gamma} = \{i \in \{1, \dots, N\} : \Gamma^i \text{ is not independent of } \mathcal{F}_{i-1}\},$$

where $\mathcal{F}_0 = \mathcal{F}$ and $\mathcal{F}_i = \sigma(\Gamma^1, \dots, \Gamma^i)$ for every $i \in \{1, \dots, N\}$. There exist $p \in \mathbb{N}^*$, $q \geqslant 2p$ and $\mathfrak{c}_{3.2} > 0$, not depending on m and N, such that

$$|\mathcal{I}_{N,\Gamma}| \leqslant \mathfrak{c}_{3.2}N^{\frac{1}{2}-\frac{p}{q}}.$$

Notation. $\mathbb{E}_0(\cdot) = \mathbb{E}(\cdot)$ and, for every $i \in \{1, \dots, N\}$, $\mathbb{E}_i(\cdot) = \mathbb{E}(\cdot|\mathcal{F}_i)$.

Note that under Assumption 3.2, for N large enough,

$$\frac{|\mathcal{I}_{N,\Gamma}|^q}{N^{q/2}\log(NT)^{q/2}} \leqslant \frac{\overline{\mathfrak{c}}_{3.2}}{N^p} \quad \text{with} \quad \overline{\mathfrak{c}}_{3.2} = \mathfrak{c}_{3.2}^q. \tag{3.6}$$

Moreover, in this section,

$$\mathfrak{c}_\Lambda = \frac{1}{256T(1+q/2)}$$

and the dimension m of $\mathcal{S}_m$ fulfills the following assumption.

Assumption 3.3 The dimension m of $\mathcal{S}_m$ satisfies

$$(\mathfrak{L}(m)^{-1}(\lambda_{\min}(\Psi_{m,\gamma}) \wedge 1))^2 \geqslant \frac{1}{\overline{\mathfrak{c}}_\Lambda} \cdot \frac{\log(NT)}{NT} \quad \text{with} \quad \overline{\mathfrak{c}}_\Lambda = \frac{\mathfrak{c}_\Lambda}{2}.$$

Let us make some remarks about Assumption 3.3:

- Note that the smaller p/q, the larger $\mathcal{I}_{N,\Gamma}$ under Assumption 3.2, but the smaller the set of *authorized* m under Assumption 3.3.
- Under Assumption 3.3, $\Psi_{m,\gamma}$ is invertible (as under Assumption 3.1) and

$$(\mathfrak{L}(m)(\|\Psi_{m,\gamma}^{-1}\|_{\text{op}} \vee 1))^2 \leqslant \overline{\mathfrak{c}}_\Lambda \frac{NT}{\log(NT)}.$$

- The set of *authorized* m is smaller under Assumption 3.3 than under Assumption 3.1. Indeed, even if $I = [\mathtt{l}, \mathtt{r}]$ with $\mathtt{l}, \mathtt{r} \in \mathbb{R}$ satisfying $\mathtt{l} < \mathtt{r}$, if there exists a constant $\underline{\mathfrak{m}} > 0$ such that $\gamma(\cdot) \geqslant \underline{\mathfrak{m}}$ on I, and if the φ_j's satisfy (3.3), then Assumption 3.3 is fulfilled by any

$$m \leqslant \frac{1}{\mathfrak{c}_\varphi^2(\underline{\mathfrak{m}}^{-1} \vee 1)} \left(\overline{\mathfrak{c}}_\Lambda \frac{NT}{\log(NT)}\right)^{\frac{1}{2}}.$$

Note that when $\Gamma^1, \dots, \Gamma^N$ have a *sparse* dependence structure, Assumption 3.1 needs to be replaced by Assumption 3.3 in order to establish a suitable bound on $\mathbb{P}(\Omega_{m,\Gamma}^c)$ by using the matrix Azuma inequality (see Proposition 3.4).

The following proposition provides a suitable bound on $\mathbb{P}(\Omega_{m,\Gamma}^c)$ under Assumptions 3.2 and 3.3.

Proposition 3.6 *Under Assumptions 3.2 and 3.3, there exists a constant* $\mathfrak{c}_{3.6} > 0$, *not depending on* m *and* N, *such that*

$$\mathbb{P}(\Omega_{m,\Gamma}^c) \leqslant \frac{\mathfrak{c}_{3.6}}{N^p}.$$

Proof First of all, note that

$$\|\widehat{G}_{m,\Gamma} - I\|_{\text{op}} \leqslant M_N + R_N \tag{3.7}$$

where

$$M_N = \left\|\sum_{i=1}^{N}(G_{m,\gamma}(\Gamma^i) - \mathbb{E}_{i-1}(G_{m,\gamma}(\Gamma^i)))\right\|_{\text{op}}$$

$$\text{and} \quad R_N = \left\|\sum_{i=1}^{N}(\mathbb{E}_{i-1}(G_{m,\gamma}(\Gamma^i)) - N^{-1}I)\right\|_{\text{op}}.$$

The proof of Proposition 3.6 is dissected in three steps. For any $\delta > 0$, the first (resp. the second) step deals with a suitable bound on $\mathbb{P}(M_N > \delta)$ (resp. $\mathbb{P}(R_N > \delta)$), and the conclusion comes in Step 3 thanks to Inequality (3.7).
Step 1. For every $i \in \{1, \dots, N\}$, since

$$G_{m,\gamma}^{\dagger}(\Gamma^i) = G_{m,\gamma}(\Gamma^i) - \mathbb{E}_{i-1}(G_{m,\gamma}(\Gamma^i))$$

is a symmetric matrix, by Proposition 3.1.(3,4), Jensen's inequality and Inequality (3.5),

$$\begin{aligned}(-G_{m,\gamma}^{\dagger}(\Gamma^i))^2 = G_{m,\gamma}^{\dagger}(\Gamma^i)^2 \preccurlyeq \lambda_{\max}(G_{m,\gamma}^{\dagger}(\Gamma^i)^2)I = \|G_{m,\gamma}(\Gamma^i) - \mathbb{E}_{i-1}(G_{m,\gamma}(\Gamma^i))\|_{\text{op}}^2 I \\ \preccurlyeq (\|G_{m,\gamma}(\Gamma^i)\|_{\text{op}} + \mathbb{E}_{i-1}(\|G_{m,\gamma}(\Gamma^i)\|_{\text{op}}))^2 I \preccurlyeq M^2\end{aligned}$$

with

$$M^2 = \frac{4}{N^2}(\mathfrak{L}(m)(\|\Psi_{m,\gamma}^{-1}\|_{\text{op}} \vee 1))^2 I.$$

So, by the matrix Azuma inequality (see Proposition 3.4),

$$\mathbb{P}\left(\lambda_{\max}\left(\sum_{i=1}^{N} G_{m,\gamma}^{\dagger}(\Gamma^i)\right) > \delta\right) \leqslant m e^{-\frac{\delta^2}{8\sigma^2}}$$

and

$$\mathbb{P}\left(-\lambda_{\min}\left(\sum_{i=1}^{N} G_{m,\gamma}^{\dagger}(\Gamma^i)\right) > \delta\right) = \mathbb{P}\left(\lambda_{\max}\left(\sum_{i=1}^{N}(-G_{m,\gamma}^{\dagger}(\Gamma^i))\right) > \delta\right) \leqslant m e^{-\frac{\delta^2}{8\sigma^2}},$$

where

$$\sigma^2 = N\|M^2\|_{\text{op}} = \frac{4}{N}(\mathfrak{L}(m)(\|\Psi_{m,\gamma}^{-1}\|_{\text{op}} \vee 1))^2.$$

By Proposition 3.1.(4), this leads to

$$\mathbb{P}(M_N > \delta) = \mathbb{P}\left(\max\left\{\lambda_{\max}\left(\sum_{i=1}^{N} G^{\dagger}_{\gamma,m}(\Gamma^i)\right); -\lambda_{\min}\left(\sum_{i=1}^{N} G^{\dagger}_{\gamma,m}(\Gamma^i)\right)\right\} > \delta\right)$$
$$\leqslant 2me^{-\frac{\delta^2}{8\sigma^2}} = 2\,m\exp\left(-\frac{\delta^2 N}{32(\mathfrak{L}(m)(\|\Psi_{m,\gamma}^{-1}\|_{\text{op}} \vee 1))^2}\right).$$

Step 2. Since $(\overline{\varphi}_1, \dots, \overline{\varphi}_m)$ is an orthonormal basis of $\mathcal{S}_m$ in $\mathbb{L}^2(\mathbb{R}, \gamma(x)dx)$ by Lemma 3.2, for every $i \in \{1, \dots, N\}\backslash\mathcal{I}_{N,\Gamma}$,

$$\mathbb{E}_{i-1}(G_{m,\gamma}(\Gamma^i)) = \mathbb{E}(G_{m,\gamma}(\Gamma)) = \left(\frac{1}{NT}\int_0^T \mathbb{E}(\overline{\varphi}_j(\Gamma_s)\overline{\varphi}_\ell(\Gamma_s))ds\right)_{j,\ell}$$
$$= \frac{1}{N}(\langle\overline{\varphi}_j, \overline{\varphi}_\ell\rangle_\gamma)_{j,\ell} = \frac{1}{N}I.$$

Then,

$$R_N = \left\|\sum_{i\in\mathcal{I}_{N,\Gamma}}(\mathbb{E}_{i-1}(G_{m,\gamma}(\Gamma^i)) - N^{-1}I)\right\|_{\text{op}}.$$

By Markov's and Jensen's inequalities, and by Inequality (3.5),

$$\mathbb{P}(R_N > \delta) \leqslant \frac{\mathbb{E}(R_N^q)}{\delta^q} \leqslant \frac{|\mathcal{I}_{N,\Gamma}|^{q-1}}{\delta^q}\sum_{i\in\mathcal{I}_{N,\Gamma}}\mathbb{E}(\|\mathbb{E}_{i-1}(G_{m,\gamma}(\Gamma^i) - N^{-1}I)\|_{\text{op}}^q)$$
$$\leqslant \frac{|\mathcal{I}_{N,\Gamma}|^q}{\delta^q}\mathbb{E}(\|G_{m,\gamma}(\Gamma) - N^{-1}I\|_{\text{op}}^q) \leqslant \frac{2^q|\mathcal{I}_{N,\Gamma}|^q}{\delta^q N^q}(\mathfrak{L}(m)(\|\Psi_{m,\gamma}^{-1}\|_{\text{op}} \vee 1))^q.$$

Step 3 (conclusion). By the two previous steps,

$$\mathbb{P}\left(\|\widehat{G}_{m,\Gamma} - I\|_{\text{op}} > \frac{1}{2}\right) \leqslant \mathbb{P}\left(\left\{M_N > \frac{1}{4}\right\} \cup \left\{R_N > \frac{1}{4}\right\}\right)$$
$$\leqslant 2\,m\exp\left[-\frac{N}{512(\mathfrak{L}(m)(\|\Psi_{m,\gamma}^{-1}\|_{\text{op}} \vee 1))^2}\right]$$
$$+ \frac{8^q|\mathcal{I}_{N,\Gamma}|^q}{N^q}(\mathfrak{L}(m)(\|\Psi_{m,\gamma}^{-1}\|_{\text{op}} \vee 1))^q.$$

Then, by Lemma 3.1, by Assumptions 3.2 and 3.3, and since $q > 2p$,

$$\mathbb{P}(\Omega_{m,\Gamma}^c) \leqslant 2\,m\exp\left(-\frac{\log(NT)}{256\mathfrak{c}_\Lambda T}\right) + \frac{8^q(\mathfrak{c}_\Lambda T)^{q/2}}{2^{q/2}} \cdot \frac{|\mathcal{I}_{N,\Gamma}|^q}{N^{q/2}\log(NT)^{q/2}}$$

$$\leqslant \mathfrak{c}_1\left(\frac{m}{N^{1+q/2}}+\frac{|\mathcal{I}_{N,\Gamma}|^q}{N^{q/2}\log(NT)^{q/2}}\right)\leqslant\frac{\mathfrak{c}_1(1+\overline{\mathfrak{c}}_{3.2})}{N^p},$$

where $\mathfrak{c}_1 > 0$ is a constant not depending on m and N. □

3.4 Is the Truncated Estimator Consistent?

For that, the event Λ_m needs to be of probability 1 when $N \to \infty$. This section deals with a suitable bound on $\mathbb{P}(\Lambda_{m,\Gamma}^c)$ inherited from the bounds on $\mathbb{P}(\Omega_{m,\Gamma}^c)$ provided by Propositions 3.5 and 3.6, where

$$\Lambda_{m,\Gamma}=\left\{\mathfrak{L}(m)(\|\widehat{\Psi}_{m,\Gamma}^{-1}\|_{\text{op}}\vee 1)\leqslant \mathfrak{c}_{\Lambda}\frac{NT}{\log(NT)}\right\}.$$

The following technical lemma has to be established first.

Lemma 3.3 *If the (positive semidefinite symmetric) matrix $\Psi_{m,\gamma}$ is invertible, then*

$$\Theta_m=\{\|\Psi_{m,\gamma}^{-1}\|_{\text{op}}<\|\widehat{\Psi}_{m,\Gamma}^{-1}-\Psi_{m,\gamma}^{-1}\|_{\text{op}}\}\subset\Omega_{m,\Gamma}^c.$$

Proof First, note that

$$\|\widehat{\Psi}_{m,\Gamma}^{-1}-\Psi_{m,\gamma}^{-1}\|_{\text{op}}=\|\Psi_{m,\gamma}^{-\frac{1}{2}}(\widehat{G}_{m,\Gamma}^{-1}-I)\Psi_{m,\gamma}^{-\frac{1}{2}}\|_{\text{op}}\leqslant\|\widehat{G}_{m,\Gamma}^{-1}-I\|_{\text{op}}\|\Psi_{m,\gamma}^{-1}\|_{\text{op}}.$$

Moreover, as established in Stewart and Sun [5], for every $A, B \in \mathcal{M}_d(\mathbb{R})$, if A is invertible and $\|A^{-1}B\|_{\text{op}} < 1$, then $M = A + B$ is invertible, and

$$\|M^{-1}-A^{-1}\|_{\text{op}}\leqslant\frac{\|B\|_{\text{op}}\|A^{-1}\|_{\text{op}}^2}{1-\|A^{-1}B\|_{\text{op}}}.$$

On $\Omega_{m,\Gamma}$, by applying this result to $A = I$ and $B = \widehat{G}_{m,\Gamma} - I$, $A + B = \widehat{G}_{m,\Gamma}$ is invertible and

$$\|\widehat{G}_{m,\Gamma}^{-1}-I\|_{\text{op}}\leqslant\frac{\|\widehat{G}_{m,\Gamma}-I\|_{\text{op}}}{1-\|\widehat{G}_{m,\Gamma}-I\|_{\text{op}}}.$$

Therefore,

$$\begin{aligned}\Theta_m&\subset\{\|\widehat{G}_{m,\Gamma}^{-1}-I\|_{\text{op}}>1\}\subset\Omega_{m,\Gamma}^c\cup(\Omega_{m,\Gamma}\cap\{\|\widehat{G}_{m,\Gamma}^{-1}-I\|_{\text{op}}>1\})\\&\subset\Omega_{m,\Gamma}^c\cup\left\{\|\widehat{G}_{m,\Gamma}-I\|_{\text{op}}\leqslant\frac{1}{2}\quad\text{and}\quad\frac{\|\widehat{G}_{m,\Gamma}-I\|_{\text{op}}}{1-\|\widehat{G}_{m,\Gamma}-I\|_{\text{op}}}>1\right\}=\Omega_{m,\Gamma}^c.\end{aligned}$$

□

Proposition 3.7 *Under Assumption 3.1 for independent* $\Gamma^1, \dots, \Gamma^N$ *(resp. under Assumptions 3.2 and 3.3),*

$$\mathbb{P}(\Lambda_{m,\Gamma}^c) \leqslant \frac{\mathfrak{c}_{3.5}}{N^p} \quad (resp.\ \mathbb{P}(\Lambda_{m,\Gamma}^c) \leqslant \frac{\mathfrak{c}_{3.6}}{N^p}).$$

Proof By Assumption 3.1 (resp. on $\Lambda_{m,\Gamma}^c$),

$$\mathfrak{L}(m)(\|\Psi_{m,\gamma}^{-1}\|_{\mathrm{op}} \vee 1) \leqslant \overline{\mathfrak{c}}_{\Lambda}\frac{NT}{\log(NT)} \quad (\text{resp. } \mathfrak{L}(m)(\|\widehat{\Psi}_{m,\Gamma}^{-1}\|_{\mathrm{op}} \vee 1) > \mathfrak{c}_{\Lambda}\frac{NT}{\log(NT)}).$$

The first inequality is equivalent to

$$\mathfrak{L}(m)\|\Psi_{m,\gamma}^{-1}\|_{\mathrm{op}} \leqslant \overline{\mathfrak{c}}_{\Lambda}\frac{NT}{\log(NT)} \quad \text{and} \quad \mathfrak{L}(m) \leqslant \overline{\mathfrak{c}}_{\Lambda}\frac{NT}{\log(NT)},$$

and then the second one leads to

$$\begin{aligned}\mathfrak{c}_{\Lambda}\frac{NT}{\log(NT)} &< \mathfrak{L}(m)\|\widehat{\Psi}_{m,\Gamma}^{-1}\|_{\mathrm{op}} \leqslant \mathfrak{L}(m)(\|\widehat{\Psi}_{m,\Gamma}^{-1} - \Psi_{m,\gamma}^{-1}\|_{\mathrm{op}} + \|\Psi_{m,\gamma}^{-1}\|_{\mathrm{op}}) \\ &\leqslant \mathfrak{L}(m)\|\widehat{\Psi}_{m,\Gamma}^{-1} - \Psi_{m,\gamma}^{-1}\|_{\mathrm{op}} + \overline{\mathfrak{c}}_{\Lambda}\frac{NT}{\log(NT)} \quad \text{on } \Lambda_{m,\Gamma}^c.\end{aligned}$$

Thus, by Lemma 3.3, by Proposition 3.5, and since $\overline{\mathfrak{c}}_{\Lambda} = \mathfrak{c}_{\Lambda}/2$,

$$\begin{aligned}\mathbb{P}(\Lambda_{m,\Gamma}^c) &\leqslant \mathbb{P}\left(\overline{\mathfrak{c}}_{\Lambda}\frac{NT}{\log(NT)} \leqslant \mathfrak{L}(m)\|\widehat{\Psi}_{m,\Gamma}^{-1} - \Psi_{m,\gamma}^{-1}\|_{\mathrm{op}}\right) \\ &\leqslant \mathbb{P}(\|\Psi_{m,\gamma}^{-1}\|_{\mathrm{op}} < \|\widehat{\Psi}_{m,\Gamma}^{-1} - \Psi_{m,\gamma}^{-1}\|_{\mathrm{op}}) \leqslant \mathbb{P}(\Omega_{m,\Gamma}^c) \leqslant \frac{\mathfrak{c}_{3.5}}{N^p}.\end{aligned}$$

The proof remains the same under Assumptions 3.2 and 3.3 by noticing that

$$\mathfrak{L}(m)(\|\Psi_{m,\gamma}^{-1}\|_{\mathrm{op}} \vee 1) \leqslant (\mathfrak{L}(m)(\|\Psi_{m,\gamma}^{-1}\|_{\mathrm{op}} \vee 1))^2 \leqslant \overline{\mathfrak{c}}_{\Lambda}\frac{NT}{\log(NT)},$$

and thanks to Proposition 3.6 instead of Proposition 3.5. □

3.5 Nonadaptive Risk Bound

By using the bounds established in Sect. 3.3 (resp. Sect. 3.4) on $\mathbb{P}(\Omega_{m,\Gamma}^c)$ (resp. $\mathbb{P}(\Lambda_{m,\Gamma}^c)$), and by using the following additional result on $\Psi_{m,\gamma}$ and $\widehat{\Psi}_{m,\Gamma}$, the purpose of this section is to establish a suitable risk bound on $\widetilde{b}_m$ for m fixed.

Lemma 3.4 *If the (positive semidefinite symmetric) matrix* $\Psi_{m,\gamma}$ *is invertible, then*

$$\widehat{\Psi}_{m,\Gamma}^{-1} \preccurlyeq 2\Psi_{m,\gamma}^{-1} \quad on \quad \Omega_{m,\Gamma} \cap \Lambda_{m,\Gamma}.$$

Proof Consider $\omega \in \Omega_{m,\Gamma} \cap \Lambda_{m,\Gamma}$. By Lemma 3.1,

$$\|\widehat{G}_{m,\Gamma}(\omega) - I\|_{\text{op}} = \sup_{x:\|x\|_{2,m}=1} |x^*(\widehat{G}_{m,\Gamma}(\omega) - I)x| \leqslant \frac{1}{2}.$$

For every $x \in \mathbb{R}^m$ such that $\|x\|_{2,m} = 1$, this leads to

$$x^*(\widehat{G}_{m,\Gamma}(\omega) - I)x \leqslant \frac{1}{2} \quad \text{and} \quad x^*(I - \widehat{G}_{m,\Gamma}(\omega))x \leqslant \frac{1}{2}.$$

So,

$$\frac{1}{2}I \preccurlyeq \widehat{G}_{m,\Gamma}(\omega) \preccurlyeq \frac{3}{2}I$$

and then, by Proposition 3.1.(3,5), $\widehat{G}_{m,\Gamma}(\omega)^{-1} \preccurlyeq 2I$. Therefore, by Proposition 3.1.(1),

$$\widehat{\Psi}_{m,\Gamma}(\omega)^{-1} = \Psi_{m,\gamma}^{-\frac{1}{2}}\widehat{G}_{m,\Gamma}(\omega)^{-1}\Psi_{m,\gamma}^{-\frac{1}{2}} \preccurlyeq 2\Psi_{m,\gamma}^{-1}.$$

□

In the sequel, $\Gamma = X$, $\gamma = f$, and the dependence in Γ and γ of operators and sets no longer appears in their notation. Moreover,

$$X^i = \mathcal{I}(x_0, W^i); \forall i \in \{1, \dots, N\},$$

where $\mathcal{I}$ is the solution map for Equation (1.1), $W^1, \dots, W^N$ are N copies of W such that

$$\mathbb{E}(W_s^i W_t^k) = R_{i,k}(s \wedge t); \forall s, t \in [0, T], \forall i, k \in \{1, \dots, N\},$$

and R is a $N \times N$ correlation matrix. From continuous-time observations, to determine the matrix R is not a statistical problem. Indeed, since σ satisfies (1.3), for every $i, k \in \{1, \dots, N\}$,

$$R_{i,k} = \frac{\langle X^i, X^k \rangle_T}{\displaystyle\int_0^T \sigma(X_s^i)\sigma(X_s^k)ds}.$$

Let us provide two examples of correlation matrix R such that Assumption 3.2 is fulfilled by $X^1, \dots, X^N$.

Example 3.1 Assume that $N \in n\mathbb{N}^*$ with $n \in \mathbb{N}^*$, and that

$$R = \begin{pmatrix} R_1 & & (0) \\ & \ddots & \\ (0) & & R_{\frac{N}{n}} \end{pmatrix},$$

where $R_1, \dots, R_{N/n}$ are N/n correlation matrices of size $n \times n$. For instance, if the number of R_i's not equal to I is of order lower than $N^{1/2-p/q}$ with $p \in \mathbb{N}^*$ and $q \geqslant 2p$, then the matrix R fulfills Assumption 3.2. For $n = 2$,

$$R_i = \begin{pmatrix} 1 & \rho_i \\ \rho_i & 1 \end{pmatrix} \quad \text{with} \quad \rho_i \in [-1, 1]; \forall i \in \left\{1, \dots, \frac{N}{2}\right\}.$$

In this special case, R fulfills Assumption 3.2 if and only if the number of non-zero ρ_i's is of order lower than $N^{1/2-p/q}$ with $p \in \mathbb{N}^*$ and $q \geqslant 2p$.

Example 3.2 Assume that $N \in 2\mathbb{N}^*$, and that

$$R = I + \begin{pmatrix} (0) & (0) & Q^* \\ (0) & (0) & (0) \\ Q & (0) & (0) \end{pmatrix},$$

where Q is a correlation matrix of size $n \times n$ with $n \in \{1, \dots, N/2\}$. If $n = n(N)$ is of order lower than $N^{1/2-p/q}$ with $p \in \mathbb{N}^*$ and $q \geqslant 2p$, then the matrix R fulfills Assumption 3.2.

Now, let us introduce two additional empirical maps:

- The empirical process ν_N, defined by

$$\nu_N(\tau) := \frac{1}{NT} \sum_{i=1}^{N} \int_0^T \tau(X_s^i)\sigma(X_s^i)dW_s^i.$$

  Note that

$$[\widehat{Z}_m]_j = \langle b_0, \varphi_j \rangle_N + \nu_N(\varphi_j); \forall j \in \{1, \dots, m\}. \tag{3.8}$$

- The empirical orthogonal projection $\widehat{\Pi}_m$ from $\mathbb{L}^2(\mathbb{R}, f(x)dx)$ onto $\mathcal{S}_m$, defined by

$$\widehat{\Pi}_m(\cdot) \in \arg\min_{\tau \in \mathcal{S}_m} \|\tau - \cdot\|_N^2. \tag{3.9}$$

Consider $h \in \mathbb{L}^2(\mathbb{R}, f(x)dx)$. For $\tau = \sum_{j=1}^m \pi_j \varphi_j$ with $\pi = (\pi_1, \dots, \pi_m) \in \mathbb{R}^m$,

$$\nabla_\tau \|\tau - h\|_N^2 = 2(\widehat{\Psi}_m \pi - (\langle h, \varphi_j \rangle_N)_j).$$

On Λ_m, since $\widehat{\Psi}_m$ is a positive definite symmetric matrix, the minimization problem (3.9) has a unique solution:

$$\widehat{\Pi}_m(h) = \sum_{j=1}^{m} [\widehat{\Psi}_m^{-1} \widehat{P}_m(h)]_j \varphi_j$$

$$\text{with} \quad \widehat{P}_m(h) = (\langle h, \varphi_j \rangle_N)_{j \in \{1,\dots,m\}}.$$

The following theorem, which is the main result of this section, provides a risk bound on $\widetilde{b}_m$.

Theorem 3.1 *Consider $b_0^I = b_0 \mathbf{1}_I$. If the matrix Ψ_m is invertible, then there exists a constant $\mathfrak{c}_{3.1,1} > 0$, not depending on m and N, such that*

$$\mathbb{E}(\|\widetilde{b}_m - b_0^I\|_N^2) \leqslant \min_{\tau \in \mathcal{S}_m} \|\tau - b_0^I\|_f^2 + \mathfrak{c}_{3.1,1} m \left(\frac{1}{N} + \mathbb{P}(\Omega_m^c)^{\frac{1}{2}} + \mathbb{P}(\Lambda_m^c)^{\frac{1}{2}} \right) \left(1 + \frac{1}{N} \sum_{l \neq k} |R_{l,k}| \right).$$

Under Assumption 3.1 for independent $W^1, \dots, W^N$ with $p \geqslant 2$, or under Assumptions 3.2 (with $p \geqslant 2$) and 3.3, there exists a constant $\mathfrak{c}_{3.1,2} > 0$, not depending on m and N, such that

$$\mathbb{E}(\|\widetilde{b}_m - b_0^I\|_N^2) \leqslant \min_{\tau \in \mathcal{S}_m} \|\tau - b_0^I\|_f^2 + \mathfrak{c}_{3.1,2} \frac{m}{N} \left(1 + \frac{1}{N} \sum_{l \neq k} |R_{l,k}| \right). \tag{3.10}$$

Proof First of all, note that

$$\|\widetilde{b}_m - b_0^I\|_N^2 = \|\widehat{b}_m - b_0^I\|_N^2 \mathbf{1}_{\Lambda_m} + \|b_0^I\|_N^2 \mathbf{1}_{\Lambda_m^c}.$$

Clearly,

$$\mathbb{E}(\|b_0^I\|_N^2 \mathbf{1}_{\Lambda_m^c}) \leqslant \mathbb{E}\left[\left(\frac{1}{NT} \sum_{i=1}^{N} \int_0^T b_0^I(X_s^i)^2 ds \right)^2 \right]^{\frac{1}{2}} \mathbb{P}(\Lambda_m^c)^{\frac{1}{2}}$$

$$\leqslant \mathfrak{c}_1 \mathbb{P}(\Lambda_m^c)^{\frac{1}{2}} \quad \text{with} \quad \mathfrak{c}_1 = \left(\int_I b_0(x)^4 f(x) dx \right)^{\frac{1}{2}}$$

and, by the definition of $\widehat{\Pi}_m$ (see (3.9)),

$$\|\widehat{b}_m - b_0^I\|_N^2 = \min_{\tau \in \mathcal{S}_m} \|\tau - b_0^I\|_N^2 + \|\widehat{b}_m - \widehat{\Pi}_m(b_0^I)\|_N^2.$$

The proof of Theorem 3.1 is dissected in four steps. The first step provides a preliminary bound on $\mathbb{E}(\|\widehat{b}_m - \widehat{\Pi}_m(b_0^I)\|_N^2 \mathbf{1}_{\Lambda_m})$, depending on

$$\mathbb{E}\left[\left(\sum_{j=1}^{m}\nu_N(\varphi_j)^2\right)^2\right] \quad \text{and} \quad \text{trace}(\Psi_m^{-1}\Phi_{m,\sigma}),$$

which are well controlled in Steps 2 and 3 respectively. The conclusion comes in Step 4 thanks to the bounds established in Sect. 3.3 (resp. Sect. 3.4) on $\mathbb{P}(\Omega_m^c)$ (resp. $\mathbb{P}(\Lambda_m^c)$).

Step 1. On Λ_m, by the decompositions of $\widehat{b}_m$ and $\widehat{\Pi}_m(b_0^I)$ in the basis $(\varphi_1,\dots,\varphi_m)$ of $\mathcal{S}_m$, and by (3.8),

$$\begin{aligned}\widehat{b}_m(\cdot)-\widehat{\Pi}_m(b_0^I)(\cdot) &= \langle\widehat{\Psi}_m^{-1}(\widehat{Z}_m-\widehat{P}_m(b_0^I)),(\varphi_j(\cdot))_j\rangle_{2,m}\\ &= \langle\widehat{\Psi}_m^{-1}\widehat{\Delta}_m,(\varphi_j(\cdot))_j\rangle_{2,m} \quad\text{with}\quad \widehat{\Delta}_m=(\nu_N(\varphi_j))_{j\in\{1,\dots,m\}}.\end{aligned}$$

Then,

$$\begin{aligned}\|\widehat{b}_m-\widehat{\Pi}_m(b_0^I)\|_N^2 &= \frac{1}{NT}\sum_{i=1}^{N}\int_0^T\left(\sum_{j=1}^{m}[\widehat{\Psi}_m^{-1}\widehat{\Delta}_m]_j\varphi_j(X_s^i)\right)^2 ds\\ &= \sum_{j,\ell=1}^{m}[\widehat{\Psi}_m^{-1}\widehat{\Delta}_m]_j[\widehat{\Psi}_m^{-1}\widehat{\Delta}_m]_\ell\langle\varphi_j,\varphi_\ell\rangle_N=\widehat{\Delta}_m^*\widehat{\Psi}_m^{-1}\widehat{\Delta}_m.\end{aligned}$$

Now, let us establish preliminary bounds on

$$\mathbb{E}(\widehat{\Delta}_m^*\widehat{\Psi}_m^{-1}\widehat{\Delta}_m\mathbf{1}_{\Lambda_m\cap\Omega_m}) \quad\text{and}\quad \mathbb{E}(\widehat{\Delta}_m^*\widehat{\Psi}_m^{-1}\widehat{\Delta}_m\mathbf{1}_{\Lambda_m\cap\Omega_m^c}).$$

On the one hand, by Lemma 3.4, on $\Lambda_m\cap\Omega_m$,

$$\widehat{\Delta}_m^*\widehat{\Psi}_m^{-1}\widehat{\Delta}_m \leqslant 2\widehat{\Delta}_m^*\Psi_m^{-1}\widehat{\Delta}_m.$$

Then,

$$\begin{aligned}\mathbb{E}(\widehat{\Delta}_m^*\widehat{\Psi}_m^{-1}\widehat{\Delta}_m\mathbf{1}_{\Lambda_m\cap\Omega_m}) &\leqslant 2\sum_{j,\ell=1}^{m}\mathbb{E}([\widehat{\Delta}_m]_j[\widehat{\Delta}_m]_\ell)[\Psi_m^{-1}]_{j,\ell}\\ &= \frac{2}{NT}\text{trace}(\Psi_m^{-1}\Phi_{m,\sigma}) \quad\text{with}\quad \Phi_{m,\sigma}=NT\mathbb{E}(\widehat{\Delta}_m\widehat{\Delta}_m^*).\end{aligned}$$

On the other hand, on Λ_m,

$$\begin{aligned}\widehat{\Delta}_m^*\widehat{\Psi}_m^{-1}\widehat{\Delta}_m &= \|\widehat{\Psi}_m^{-\frac{1}{2}}\widehat{\Delta}_m\|_{2,m}^2\\ &\leqslant \|\widehat{\Psi}_m^{-1}\|_{\text{op}}|\widehat{\Delta}_m^*\widehat{\Delta}_m| \leqslant \mathfrak{c}_\Lambda\frac{NT}{\log(NT)\mathfrak{L}(m)}\sum_{j=1}^{m}\nu_N(\varphi_j)^2.\end{aligned}$$

Then,

$$\mathbb{E}(\widehat{\Delta}_m^*\widehat{\Psi}_m^{-1}\widehat{\Delta}_m\mathbf{1}_{\Lambda_m\cap\Omega_m^c}) \leqslant \mathfrak{c}_\Lambda\frac{NT}{\log(NT)\mathfrak{L}(m)}\mathbb{E}\left[\left(\sum_{j=1}^m \nu_N(\varphi_j)^2\right)^2\right]^{\frac{1}{2}}\mathbb{P}(\Omega_m^c)^{\frac{1}{2}}.$$

Therefore, to conclude this first step,

$$\mathbb{E}(\|\widehat{b}_m - \widehat{\Pi}_m(b_0^I)\|_N^2\mathbf{1}_{\Lambda_m}) \leqslant \frac{2}{NT}\text{trace}(\Psi_m^{-1}\Phi_{m,\sigma}) + \mathfrak{c}_\Lambda\frac{NT}{\log(NT)\mathfrak{L}(m)}\mathbb{E}\left[\left(\sum_{j=1}^m \nu_N(\varphi_j)^2\right)^2\right]^{\frac{1}{2}}\mathbb{P}(\Omega_m^c)^{\frac{1}{2}}.$$

Step 2. By Jensen's and Burkholder-Davis-Gundy's inequalities (see Sect. 3.1), and since $d\langle W^i, W^k\rangle_t = R_{i,k}dt$ for every $i, k \in \{1, \dots, N\}$,

$$\mathbb{E}\left[\left(\sum_{j=1}^m \nu_N(\varphi_j)^2\right)^2\right] \leqslant m\sum_{j=1}^m \mathbb{E}(\nu_N(\varphi_j)^4) \leqslant \mathfrak{b}_4 m\sum_{j=1}^m \mathbb{E}(\langle\nu_N(\varphi_j)\rangle_T^2)$$
$$\leqslant \frac{2\mathfrak{b}_4 m}{N^4T^4}\sum_{j=1}^m(\mathbb{E}(\delta_j^2) + \mathbb{E}(\rho_j^2))$$

where, for every $j \in \{1, \dots, m\}$,

$$\delta_j = \sum_{i=1}^N\int_0^T \varphi_j(X_s^i)^2\sigma(X_s^i)^2ds$$
$$\text{and}\quad \rho_j = \sum_{i\neq k}R_{i,k}\int_0^T \varphi_j(X_s^i)\sigma(X_s^i)\varphi_j(X_s^k)\sigma(X_s^k)ds.$$

On the one hand, by Jensen's inequality,

$$\sum_{j=1}^m \mathbb{E}(\delta_j^2) \leqslant NT\sum_{j=1}^m\sum_{i=1}^N\int_0^T \mathbb{E}(\varphi_j(X_s^i)^4\sigma(X_s^i)^4)ds$$
$$\leqslant N^2T\mathfrak{L}(m)^2\int_0^T \mathbb{E}(\sigma(X_s)^4)ds = N^2T^2\mathfrak{L}(m)^2\int_{-\infty}^{\infty}\sigma(x)^4 f(x)dx.$$

On the other hand, by Jensen's and Cauchy-Schwarz's inequalities,

$$\sum_{j=1}^{m}\mathbb{E}(\rho_j^2) \leqslant T\left(\sum_{i\neq k}|R_{i,k}|\right)\sum_{j=1}^{m}\sum_{i\neq k}|R_{i,k}|\int_0^T \mathbb{E}(\varphi_j(X_s^i)^2\sigma(X_s^i)^2\varphi_j(X_s^k)^2\sigma(X_s^k)^2)ds$$
$$\leqslant T\left(\sum_{i\neq k}|R_{i,k}|\right)^2\sum_{j=1}^{m}\int_0^T \mathbb{E}(\varphi_j(X_s)^4\sigma(X_s)^4)ds$$
$$\leqslant T^2\left(\sum_{i\neq k}|R_{i,k}|\right)^2\mathfrak{L}(m)^2\int_{-\infty}^{\infty}\sigma(x)^4 f(x)dx.$$

So,

$$\mathbb{E}\left[\left(\sum_{j=1}^{m}\nu_N(\varphi_j)^2\right)^2\right] \leqslant \mathfrak{c}_2\frac{m\mathfrak{L}(m)^2}{N^2T^2}\left[1+\left(\frac{1}{N}\sum_{i\neq k}|R_{i,k}|\right)^2\right]$$

with

$$\mathfrak{c}_2 = 2\mathfrak{b}_4\int_{-\infty}^{\infty}\sigma(x)^4 f(x)dx.$$

Therefore, there exists a constant $\mathfrak{c}_3 > 0$, not depending on m and N, such that

$$\mathfrak{c}_\Lambda\frac{NT}{\log(NT)\mathfrak{L}(m)}\mathbb{E}\left[\left(\sum_{j=1}^{m}\nu_N(\varphi_j)^2\right)^2\right]^{\frac{1}{2}} \leqslant \mathfrak{c}_3 m^{\frac{1}{2}}\left(1+\frac{1}{N}\sum_{i\neq k}|R_{i,k}|\right).$$

Step 3. First, let us show that $\Phi_{m,\sigma}$ is a positive semidefinite symmetric matrix. Indeed, for any $y \in \mathbb{R}^m$,

$$y^*\Phi_{m,\sigma}y = \frac{1}{NT}\sum_{j,\ell=1}^{m} y_j y_\ell$$
$$\times\sum_{i,k=1}^{N}\mathbb{E}\left(\left(\int_0^T \varphi_j(X_s^i)\sigma(X_s^i)dW_s^i\right)\left(\int_0^T \varphi_\ell(X_s^k)\sigma(X_s^k)dW_s^k\right)\right)$$
$$= \frac{1}{NT}\mathbb{E}\left[\left(\sum_{i=1}^{N}\int_0^T \tau_y(X_s^i)\sigma(X_s^i)dW_s^i\right)^2\right] \geqslant 0 \quad\text{with}\quad \tau_y = \sum_{j=1}^{m} y_j\varphi_j.$$

Since $d\langle W^i, W^k\rangle_t = R_{i,k}dt$ for every $i, k \in \{1,\dots,N\}$, and by the stochastic integration by parts formula,

$$y^*\Phi_{m,\sigma}y = \frac{1}{NT}\mathbb{E}\left[\left(\sum_{i=1}^{N}\int_0^T \tau_y(X_s^i)\sigma(X_s^i)dW_s^i\right)^2\right]$$

$$\leqslant \frac{1}{T}\left(1+\frac{1}{N}\sum_{i\neq k}|R_{i,k}|\right)\int_0^T \mathbb{E}(\tau_y(X_s)^2\sigma(X_s)^2)ds$$

$$\leqslant \left(1+\frac{1}{N}\sum_{i\neq k}|R_{i,k}|\right)$$

$$\times \int_{-\infty}^{\infty}\left(\sum_{j=1}^{m} y_j\varphi_j(x)\right)^2\sigma(x)^2 f(x)dx$$

$$\leqslant \|\sigma\|_\infty^2\left(1+\frac{1}{N}\sum_{i\neq k}|R_{i,k}|\right)\|\Psi_m^{\frac{1}{2}}y\|_{2,m}^2. \tag{3.11}$$

Therefore, since $\Phi_{m,\sigma}$ is a positive semidefinite symmetric matrix, by Proposition 3.1.(4,6) and Inequality (3.11),

$$\frac{1}{m}\mathrm{trace}(\Psi_m^{-1}\Phi_{m,\sigma}) \leqslant \|\Psi_m^{-\frac{1}{2}}\Phi_{m,\sigma}\Psi_m^{-\frac{1}{2}}\|_{\mathrm{op}}$$

$$= \sup_{y:\|\Psi_m^{1/2}y\|_{2,m}=1} y^*\Phi_{m,\sigma}y \leqslant \|\sigma\|_\infty^2\left(1+\frac{1}{N}\sum_{i\neq k}|R_{i,k}|\right).$$

Step 4 (conclusion). By the three previous steps,

$$\mathbb{E}(\|\widetilde{b}_m - b_0^I\|_N^2) \leqslant \min_{\tau\in\mathcal{S}_m}\|\tau - b_0^I\|_f^2 + \mathfrak{c}_1\mathbb{P}(\Lambda_m^c)^{\frac{1}{2}}$$

$$+\left(\frac{2m}{NT}\|\sigma\|_\infty^2 + \mathfrak{c}_3 m^{\frac{1}{2}}\mathbb{P}(\Omega_m^c)^{\frac{1}{2}}\right)\left(1+\frac{1}{N}\sum_{i\neq k}|R_{i,k}|\right).$$

Under Assumption 3.1 for independent $W^1,\dots,W^N$ with $p\geqslant 2$ (resp. under Assumptions 3.2 (with $p\geqslant 2$) and 3.3), by Propositions 3.5 (resp. 3.6) and 3.7, there exists a constant $\mathfrak{c}_4>0$, not depending on m and N, such that

$$\mathbb{E}(\|\widetilde{b}_m - b_0^I\|_N^2) \leqslant \min_{\tau\in\mathcal{S}_m}\|\tau - b_0^I\|_f^2 + \mathfrak{c}_4\frac{m}{N}\left(1+\frac{1}{N}\sum_{i\neq k}|R_{i,k}|\right).$$

□

Let us conclude this section with some remarks about Theorem 3.1:

- Note that

$$\mathcal{I}_N = \{i\in\{2,\dots,N\} : \exists k\in\{1,\dots,i-1\} \text{ such that } R_{i,k}\neq 0\}.$$

So, in the statement of Theorem 3.1, Assumption 3.2 concerns the matrix R directly, which legitimates to say that $W^1, \dots, W^N$ (and then $X^1, \dots, X^N$) have a *sparse* dependence structure.

- Note that the *variance term* in Inequality (3.10) is of order

$$\frac{m}{N}\left(1 + \frac{1}{N}\sum_{i \neq k} |R_{i,k}|\right),$$

which becomes m/N, as in the nonparametric regression framework, when $W^1, \dots, W^N$ are independent.

- The order of the *bias term*

$$\min_{\tau \in \mathcal{S}_m} \|\tau - b_0^I\|_f^2 \quad \text{in Inequality (3.10) depends on the } \varphi_j\text{'s.}$$

For instance, assume that $I = [\mathtt{l}, \mathtt{r}]$ with $\mathtt{l}, \mathtt{r} \in \mathbb{R}$ satisfying $\mathtt{l} < \mathtt{r}$, and that $(\varphi_1, \dots, \varphi_m)$ is the I-supported trigonometric basis. Consider $\beta \in \mathbb{N}^*$, the Fourier-Sobolev space

$$\mathbb{W}_2^{\beta}([\mathtt{l}, \mathtt{r}]) = \left\{\varphi : [\mathtt{l}, \mathtt{r}] \rightarrow \mathbb{R}\ \beta \text{ times differentiable: } \int_{\mathtt{l}}^{\mathtt{r}} \varphi^{(\beta)}(x)^2 dx < \infty\right\},$$

and assume that $(b_0)_{|I} \in \mathbb{W}_0^{\beta}([\mathtt{l}, \mathtt{r}])$. By DeVore and Lorentz [6], Corollary 2.4 p. 205, there exists a constant $\mathfrak{c}_{\beta,\mathtt{l},\mathtt{r}} > 0$, not depending on m, such that

$$\|\Pi_m(b_0^I) - b_0^I\|^2 \leqslant \mathfrak{c}_{\beta,\mathtt{l},\mathtt{r}} m^{-2\beta},$$

where Π_m is the orthogonal projection from $\mathbb{L}^2(\mathbb{R}, dx)$ onto $\mathcal{S}_m$. Since f is upper bounded on I by a constant $\overline{\mathfrak{m}} > 0$ thanks to Inequality (3.1),

$$\begin{aligned}\min_{\tau \in \mathcal{S}_m} \|\tau - b_0^I\|_f^2 &\leqslant \overline{\mathfrak{m}} \|\Pi_m(b_0^I) - b_0^I\|^2 \\ &\leqslant \overline{\mathfrak{c}}_{\beta,\mathtt{l},\mathtt{r}} m^{-2\beta} \quad \text{with} \quad \overline{\mathfrak{c}}_{\beta,\mathtt{l},\mathtt{r}} = \mathfrak{c}_{\beta,\mathtt{l},\mathtt{r}} \overline{\mathfrak{m}}.\end{aligned}$$

In conclusion, by Theorem 3.1, there exists a constant $\overline{\mathfrak{c}}_{3.1,1} > 0$, not depending on m and N, such that

$$\mathbb{E}(\|\widetilde{b}_m - b_0^I\|_N^2) \leqslant \overline{\mathfrak{c}}_{3.1,1}\left(m^{-2\beta} + \frac{m}{N}\left(1 + \frac{1}{N}\sum_{i \neq k} |R_{i,k}|\right)\right),$$

and then the bias-variance tradeoff is reached by (the risk bound on) $\widetilde{b}_m$ for m of order

$$\left(\frac{1}{N}\left(1+\frac{1}{N}\sum_{i\neq k}|R_{i,k}|\right)\right)^{-\frac{1}{1+2\beta}}.$$

3.6 Model Selection

In practice, the dimension m of $\mathcal{S}_m$ needs to be selected from data because the one for which $\widetilde{b}_m$ reaches the bias-variance tradeoff depends on some unknown regularity parameters of b_0. In order to introduce an appropriate model selection criterion, throughout this section, the φ_j's and the matrix R fulfill the following additional assumptions.

Assumption 3.4 The φ_j's satisfy the two following conditions:

1. For every $m, m' \in \{1, \dots, N_T\}$, if $m > m'$, then $\mathcal{S}_{m'} \subset \mathcal{S}_m$.
2. There exists a constant $\mathfrak{c}_\varphi \geqslant 1$, not depending on N, such that for every $m \in \{1, \dots, N_T\}$, $\mathfrak{L}(m) \leqslant \mathfrak{c}_\varphi^2 m$.

Assumption 3.5 There exists a constant $\mathfrak{m}_{3.5} > 0$, not depending on N, such that $\|R\|_{\text{op}} \leqslant \mathfrak{m}_{3.5}$.

Since R is a symmetric matrix, there exist an orthogonal matrix P and a diagonal matrix D such that $R = PDP^*$. Then,

$$\|R\|_{\text{op}} = \|D\|_{\text{op}} = \sup_{\lambda \in \text{sp}(R)} |\lambda|.$$

So, the matrix R fulfills Assumption 3.5 if and only if there exists a constant $\mathfrak{m} > 0$, not depending on N, such that $|\lambda| \leqslant \mathfrak{m}$ for every $\lambda \in \text{sp}(R)$. Let us provide two examples of correlation matrices fulfilling both Assumptions 3.2 and 3.5.

Example 3.3 The notations are those of Example 3.1. First, assume that $n = 2$. For every $\lambda \in \mathbb{R}$,

$$\det(R - \lambda I) = \prod_{i=1}^{\frac{N}{2}} \det(R_i - \lambda I) = \prod_{i=1}^{\frac{N}{2}} (1 - \lambda - \rho_i)(1 - \lambda + \rho_i),$$

and then

$$\text{sp}(R) = \left\{1 \pm \rho_i; i = 1, \dots, \frac{N}{2}\right\}.$$

So, the matrix R fulfills Assumption 3.5:

$$\|R\|_{\text{op}} = \max_{i\in\{1,\dots,N/2\}} |1 \pm \rho_i| \leqslant 2.$$

More generally, assume that $n \geqslant 2$. For every $\lambda \in \mathbb{R}$,

$$\det(R - \lambda I) = \prod_{i=1}^{\frac{N}{n}} \det(R_i - \lambda I),$$

and then

$$\text{sp}(R) = \bigcup_{i=1}^{\frac{N}{n}} \text{sp}(R_i).$$

So, the matrix R fulfills Assumption 3.5:

$$\begin{aligned}\|R\|_{\text{op}} &= \max_{i\in\{1,\dots,N/n\}} \left\{\sup_{\lambda\in\text{sp}(R_i)} |\lambda|\right\}\\ &= \max_{i\in\{1,\dots,N/n\}} \|R_i\|_{\text{op}} \leqslant \max_{i\in\{1,\dots,N/n\}} \left(\sum_{k,k'=1}^{n} [R_i]_{k,k'}^2\right)^{\frac{1}{2}} \leqslant n.\end{aligned}$$

Example 3.4 The notations are those of Example 3.2. Let us show that the matrix R fulfills Assumption 3.5 if and only if

$$\sup\{|\lambda|;\ \lambda \in \mathbb{R} \text{ such that } q(\lambda) = \det((1-\lambda)^2 I - Q^* Q) = 0\}$$
$$\text{is bounded by a constant not depending on N} \tag{3.12}$$

Consider

$$\mathcal{Q} = \begin{pmatrix} (0) & (0) \\ Q & (0) \end{pmatrix} \in \mathcal{M}_{\frac{N}{2}}(\mathbb{R}),$$

and note that

$$\mathcal{Q}^*\mathcal{Q} = \begin{pmatrix} Q^*Q & (0) \\ (0) & (0) \end{pmatrix}.$$

Then, for every $\lambda \in \mathbb{R}$,

$$\begin{aligned}\det(R - \lambda I) &= \det\begin{pmatrix} (1-\lambda)I & \mathcal{Q}^* \\ \mathcal{Q} & (1-\lambda)I \end{pmatrix} = \det((1-\lambda)^2 I - \mathcal{Q}^*\mathcal{Q})\\ &= \det\begin{pmatrix} (1-\lambda)^2 I - Q^*Q & (0) \\ (0) & (1-\lambda)^2 I \end{pmatrix} = (1-\lambda)^{N-2n} q(\lambda).\end{aligned}$$

As expected, R fulfills Assumption 3.5 if and only if Q fulfills (3.12). For instance, assume that

$$Q = \begin{pmatrix} q_1 & & (0) \\ & \ddots & \\ (0) & & q_n \end{pmatrix} \quad \text{with} \quad q_1, \dots, q_n \in [-1, 1].$$

In this special case, R is a Toeplitz matrix,

$$\begin{aligned} q(\lambda) &= \det((1-\lambda)^2 I - Q^2) \\ &= \prod_{i=1}^{n} (1 - \lambda - q_i)(1 - \lambda + q_i), \end{aligned}$$

and then Q fulfills (3.12).

Now, let us consider

$$\widehat{m} = \arg \min_{m \in \widehat{\mathcal{M}}_N} \{-\|\widehat{b}_m\|_N^2 + \text{pen}(m)\} \tag{3.13}$$

where

$$\text{pen}(m) = \mathfrak{c}_{\text{cal}} \frac{m}{N} \left(1 + \frac{1}{N} \sum_{i \neq k} |R_{i,k}|\right) ; \forall m \in \{1, \dots, N_T\},$$

the constant $\mathfrak{c}_{\text{cal}} > 0$ needs to be calibrated in practice,

$$\widehat{\mathcal{M}}_N = \left\{ m \in \{1, \dots, N_T\} : (\mathfrak{c}_\varphi^2 m (\|\widehat{\Psi}_m^{-1}\|_{\text{op}} \vee 1))^2 \leqslant \mathfrak{d} \frac{NT}{\log(NT)} \right\}$$

and, under Assumption 3.2,

$$\mathfrak{d} = \frac{1}{2048 \mathfrak{c}_\varphi^4 T (1 + q/2)} \leqslant \frac{\overline{\mathfrak{c}}_\Lambda}{4}.$$

Consider also the theoretical counterpart

$$\mathcal{M}_N = \left\{ m \in \{1, \dots, N_T\} : (\mathfrak{c}_\varphi^2 m (\|\Psi_m^{-1}\|_{\text{op}} \vee 1))^2 \leqslant \overline{\mathfrak{d}} \frac{NT}{\log(NT)} \right\} \quad \text{of} \quad \widehat{\mathcal{M}}_N,$$

where $\overline{\mathfrak{d}} = \mathfrak{d}/4$.

In order to establish a risk bound on the adaptive estimator $\widehat{b}_{\widehat{m}}$, the deterministic set $\mathcal{M}_N$ needs to be almost surely contained in its empirical counterpart when $N \to \infty$. To that purpose, let us introduce the event

$$\Xi_{N,\Gamma} = \{\mathcal{M}_{N,\gamma} \subset \widehat{\mathcal{M}}_{N,\Gamma} \subset \mathfrak{M}_{N,\gamma}\},$$

where

- $\widehat{\mathcal{M}}_{N,\Gamma} = \left\{m \in \{1,\dots,N_T\} : (\mathfrak{c}_\varphi^2 m(\|\widehat{\Psi}_{m,\Gamma}^{-1}\|_{\text{op}} \vee 1))^2 \leqslant \mathfrak{d}\frac{NT}{\log(NT)}\right\}$,
- $\mathcal{M}_{N,\gamma} = \left\{m \in \{1,\dots,N_T\} : (\mathfrak{c}_\varphi^2 m(\|\Psi_{m,\gamma}^{-1}\|_{\text{op}} \vee 1))^2 \leqslant \overline{\mathfrak{d}}\frac{NT}{\log(NT)}\right\}$,
- and $\mathfrak{M}_{N,\gamma} = \left\{m \in \{1,\dots,N_T\} : (\mathfrak{c}_\varphi^2 m(\|\Psi_{m,\gamma}^{-1}\|_{\text{op}} \vee 1))^2 \leqslant \widetilde{\mathfrak{d}}\frac{NT}{\log(NT)}\right\}$ with $\widetilde{\mathfrak{d}} = 4\mathfrak{d}$.

The following proposition provides a suitable bound on $\mathbb{P}(\Xi_{N,\Gamma}^c)$ when $\Gamma^1,\dots,\Gamma^N$ have a *sparse* dependence structure (i.e., fulfill Assumption 3.2).

Proposition 3.8 *Under Assumptions 3.2 and 3.4, there exists a constant $\mathfrak{c}_{3.8} > 0$, not depending on N, such that*

$$\mathbb{P}(\Xi_{N,\Gamma}^c) \leqslant \frac{\mathfrak{c}_{3.8}}{N^{p-1}}.$$

Proof First of all, note that

$$\Xi_{N,\Gamma}^c = \{\mathcal{M}_{N,\gamma} \not\subset \widehat{\mathcal{M}}_{N,\Gamma}\} \cup \{\widehat{\mathcal{M}}_{N,\Gamma} \not\subset \mathfrak{M}_{N,\gamma}\}.$$

The proof of Proposition 3.8 is dissected in three steps. Step 1 provides a bound on $\mathbb{P}(\mathcal{M}_{N,\gamma} \not\subset \widehat{\mathcal{M}}_{N,\Gamma})$, Step 2 a bound on $\mathbb{P}(\|\widehat{\Psi}_{m,\Gamma} - \Psi_{m,\gamma}\|_{\text{op}} > \delta)$ for every $m \in \{1,\dots,N_T\}$ and $\delta > 0$, and then Step 3 a bound on $\mathbb{P}(\widehat{\mathcal{M}}_{N,\Gamma} \not\subset \mathfrak{M}_{N,\gamma})$.
Step 1. On $\{\mathcal{M}_{N,\gamma} \not\subset \widehat{\mathcal{M}}_{N,\Gamma}\}$, there exists $m \in \{1,\dots,N_T\}$ such that

$$(\mathfrak{c}_\varphi^2 m(\|\Psi_{m,\gamma}^{-1}\|_{\text{op}} \vee 1))^2 \leqslant \overline{\mathfrak{d}}\frac{NT}{\log(NT)} \quad\text{and}\quad (\mathfrak{c}_\varphi^2 m(\|\widehat{\Psi}_{m,\Gamma}^{-1}\|_{\text{op}} \vee 1))^2 > \mathfrak{d}\frac{NT}{\log(NT)}.$$

The first inequality is equivalent to

$$\mathfrak{c}_\varphi^4 m^2\|\Psi_{m,\gamma}^{-1}\|_{\text{op}}^2 \leqslant \overline{\mathfrak{d}}\frac{NT}{\log(NT)} \quad\text{and}\quad \mathfrak{c}_\varphi^4 m^2 \leqslant \overline{\mathfrak{d}}\frac{NT}{\log(NT)},$$

and then the second one leads to

$$\begin{aligned}\mathfrak{d}\frac{NT}{\log(NT)} < \mathfrak{c}_\varphi^4 m^2\|\widehat{\Psi}_{m,\Gamma}^{-1}\|_{\text{op}}^2 &\leqslant 2\mathfrak{c}_\varphi^4 m^2(\|\widehat{\Psi}_{m,\Gamma}^{-1} - \Psi_{m,\gamma}^{-1}\|_{\text{op}}^2 + \|\Psi_{m,\gamma}^{-1}\|_{\text{op}}^2)\\ &\leqslant 2\mathfrak{c}_\varphi^4 m^2\|\widehat{\Psi}_{m,\Gamma}^{-1} - \Psi_{m,\gamma}^{-1}\|_{\text{op}}^2 + 2\overline{\mathfrak{d}}\frac{NT}{\log(NT)}.\end{aligned}$$

So, since $\mathfrak{d} - 2\overline{\mathfrak{d}} = 2\overline{\mathfrak{d}}$, and by the definition of $\mathcal{M}_{N,\gamma}$,

$$\{\mathcal{M}_{N,\gamma} \not\subset \widehat{\mathcal{M}}_{N,\Gamma}\} \subset \bigcup_{m\in\mathcal{M}_{N,\gamma}} \left\{\overline{\mathfrak{d}}\frac{NT}{\log(NT)} \leqslant \mathfrak{c}_\varphi^4 m^2\|\widehat{\Psi}_{m,\Gamma}^{-1} - \Psi_{m,\gamma}^{-1}\|_{\text{op}}^2\right\}$$

$$\subset \bigcup_{m\in\mathcal{M}_{N,\gamma}} \{\|\Psi_{m,\gamma}^{-1}\|_{\text{op}} < \|\widehat{\Psi}_{m,\Gamma}^{-1} - \Psi_{m,\gamma}^{-1}\|_{\text{op}}\} \subset \bigcup_{m\in\mathcal{M}_{N,\gamma}} \Omega_{m,\Gamma}^{c}.$$

Therefore, since $\overline{\mathfrak{d}} \leqslant \overline{\mathfrak{c}}_{\Lambda}$, by Proposition 3.6,

$$\mathbb{P}(\mathcal{M}_{N,\gamma} \not\subset \widehat{\mathcal{M}}_{N,\Gamma}) \leqslant \sum_{m\in\mathcal{M}_{N,\gamma}} \mathbb{P}(\Omega_{m,\Gamma}^{c}) \leqslant \frac{\mathfrak{c}_1}{N^{p-1}},$$

where $\mathfrak{c}_1 > 0$ is a constant not depending on N.

Step 2. For any $m \in \{1, \dots, N_T\}$, note that

$$\|\widehat{\Psi}_{m,\Gamma} - \Psi_{m,\gamma}\|_{\text{op}} \leqslant M_N + R_N,$$

where

$$M_N = \frac{1}{N}\left\|\sum_{i=1}^{N}(\Psi_m(\Gamma^i) - \mathbb{E}_{i-1}(\Psi_m(\Gamma^i)))\right\|_{\text{op}}$$

$$\text{and}\quad R_N = \frac{1}{N}\left\|\sum_{i=1}^{N}(\mathbb{E}_{i-1}(\Psi_m(\Gamma^i)) - \mathbb{E}(\Psi_m(\Gamma^i)))\right\|_{\text{op}}.$$

On the one hand, for any $\Delta > 0$, let us establish a suitable bound on $\mathbb{P}(M_N > \Delta)$. For every $i \in \{1, \dots, N\}$, since

$$\Psi_m^{\dagger}(\Gamma^i) = \frac{1}{N}(\Psi_m(\Gamma^i) - \mathbb{E}_{i-1}(\Psi_m(\Gamma^i)))$$

is a symmetric matrix, by Proposition 3.1.(3,4), Jensen's inequality and Inequality (3.4),

$$\begin{aligned}(-\Psi_m^{\dagger}(\Gamma^i))^2 = \Psi_m^{\dagger}(\Gamma^i)^2 \preccurlyeq \lambda_{\max}(\Psi_m^{\dagger}(\Gamma^i)^2)I &= \frac{1}{N^2}\|\Psi_m(\Gamma^i) - \mathbb{E}_{i-1}(\Psi_m(\Gamma^i))\|_{\text{op}}^2 I\\ &\preccurlyeq \frac{1}{N^2}(\|\Psi_m(\Gamma^i)\|_{\text{op}} + \mathbb{E}_{i-1}(\|\Psi_m(\Gamma^i)\|_{\text{op}}))^2 \preccurlyeq M^2\end{aligned}$$

with

$$M^2 = \frac{4\mathfrak{L}(m)^2}{N^2}I.$$

So, by the matrix Azuma inequality (see Proposition 3.4),

$$\mathbb{P}(M_N > \Delta) = \mathbb{P}\left(\left\|\sum_{i=1}^{N}\Psi_m^{\dagger}(\Gamma^i)\right\|_{\text{op}} > \Delta\right) \leqslant 2m\exp\left(-\frac{\Delta^2 N}{32\mathfrak{L}(m)^2}\right).$$

On the other hand, let us establish a suitable bound on $\mathbb{P}(R_N > \Delta)$. By the definition of $\mathcal{I}_{N,\Gamma}$,

$$R_N = \frac{1}{N}\left\|\sum_{i\in\mathcal{I}_{N,\Gamma}}(\mathbb{E}_{i-1}(\Psi_m^{\dagger}(\Gamma^i)) - \mathbb{E}(\Psi_m^{\dagger}(\Gamma^i)))\right\|_{\mathrm{op}}.$$

By Markov's and Jensen's inequalities, and by Inequality (3.4),

$$\begin{aligned}\mathbb{P}(R_N > \Delta) &\leqslant \frac{\mathbb{E}(R_N^q)}{\Delta^q} \leqslant \frac{|\mathcal{I}_{N,\Gamma}|^{q-1}}{\Delta^q N^q}\sum_{i\in\mathcal{I}_{N,\Gamma}}\mathbb{E}(\|\mathbb{E}_{i-1}(\Psi_m^{\dagger}(\Gamma^i) - \mathbb{E}(\Psi_m^{\dagger}(\Gamma^i)))\|_{\mathrm{op}}^q)\\ &\leqslant \frac{|\mathcal{I}_{N,\Gamma}|^q}{\Delta^q N^q}\mathbb{E}(\|\Psi_m^{\dagger}(\Gamma) - \mathbb{E}(\Psi_m^{\dagger}(\Gamma))\|_{\mathrm{op}}^q) \leqslant \frac{2^q|\mathcal{I}_{N,\Gamma}|^q}{\Delta^q N^q}\mathfrak{L}(m)^q.\end{aligned}$$

Therefore, for every $\delta > 0$,

$$\begin{aligned}\mathbb{P}(\|\widehat{\Psi}_{m,\Gamma} - \Psi_{m,\gamma}\|_{\mathrm{op}} > \delta) &\leqslant \mathbb{P}\left(\left\{M_N > \frac{\delta}{2}\right\}\cup\left\{R_N > \frac{\delta}{2}\right\}\right)\\ &\leqslant 2m\exp\left(-\frac{\delta^2 N}{128\mathfrak{L}(m)^2}\right) + \frac{4^q|\mathcal{I}_{N,\Gamma}|^q}{\delta^q N^q}\mathfrak{L}(m)^q.\end{aligned}$$

Step 3. On $\{\widehat{\mathcal{M}}_{N,\Gamma} \not\subset \mathfrak{M}_{N,\gamma}\}$, there exists $m \in \{1,\dots,N_T\}$ such that

$$(\mathfrak{c}_\varphi^2 m(\|\widehat{\Psi}_{m,\Gamma}^{-1}\|_{\mathrm{op}} \vee 1))^2 \leqslant \mathfrak{d}\frac{NT}{\log(NT)} \quad\text{and}\quad (\mathfrak{c}_\varphi^2 m(\|\Psi_{m,\gamma}^{-1}\|_{\mathrm{op}} \vee 1))^2 > \widetilde{\mathfrak{d}}\frac{NT}{\log(NT)}.$$

The first inequality is equivalent to

$$\mathfrak{c}_\varphi^4 m^2\|\widehat{\Psi}_{m,\Gamma}^{-1}\|_{\mathrm{op}}^2 \leqslant \mathfrak{d}\frac{NT}{\log(NT)} \quad\text{and}\quad \mathfrak{c}_\varphi^4 m^2 \leqslant \mathfrak{d}\frac{NT}{\log(NT)},$$

and then the second one leads to

$$\begin{aligned}\widetilde{\mathfrak{d}}\frac{NT}{\log(NT)} < \mathfrak{c}_\varphi^4 m^2\|\Psi_{m,\gamma}^{-1}\|_{\mathrm{op}}^2 &\leqslant 2\mathfrak{c}_\varphi^4 m^2(\|\Psi_{m,\gamma}^{-1} - \widehat{\Psi}_{m,\Gamma}^{-1}\|_{\mathrm{op}}^2 + \|\widehat{\Psi}_{m,\Gamma}^{-1}\|_{\mathrm{op}}^2)\\ &\leqslant 2\mathfrak{c}_\varphi^4 m^2\|\Psi_{m,\gamma}^{-1} - \widehat{\Psi}_{m,\Gamma}^{-1}\|_{\mathrm{op}}^2 + 2\mathfrak{d}\frac{NT}{\log(NT)}.\end{aligned}$$

Moreover, for every $m \in \{1,\dots,N_T\}$, by interchanging $\widehat{\Psi}_{m,\Gamma}$ and $\Psi_{m,\gamma}$ in the proof of Lemma 3.3,

$$\begin{aligned}\{\|\Psi_{m,\gamma}^{-1} - \widehat{\Psi}_{m,\Gamma}^{-1}\|_{\mathrm{op}} > \|\widehat{\Psi}_{m,\Gamma}^{-1}\|_{\mathrm{op}}\} &\subset \left\{\|\widehat{\Psi}_{m,\Gamma}^{-\frac{1}{2}}\Psi_{m,\gamma}\widehat{\Psi}_{m,\Gamma}^{-\frac{1}{2}} - I\|_{\mathrm{op}} > \frac{1}{2}\right\}\\ &\subset \left\{\|\widehat{\Psi}_{m,\Gamma} - \Psi_{m,\gamma}\|_{\mathrm{op}} > \frac{1}{2}\|\widehat{\Psi}_{m,\Gamma}^{-1}\|_{\mathrm{op}}^{-1}\right\}.\end{aligned}$$

So, since $\widetilde{\mathfrak{d}} - 2\mathfrak{d} = 2\mathfrak{d}$,

$$\begin{aligned}
\{\widehat{\mathcal{M}}_{N,\Gamma} \not\subset \mathfrak{M}_{N,\gamma}\}
&\subset \bigcup_{\mathfrak{c}_{\varphi}^4 m^2 \leqslant \mathfrak{d}\frac{NT}{\log(NT)}} \left\{\mathfrak{c}_{\varphi}^4 m^2 \|\widehat{\Psi}_{m,\Gamma}^{-1}\|_{\mathrm{op}}^2 \leqslant \mathfrak{d}\frac{NT}{\log(NT)} < \mathfrak{c}_{\varphi}^4 m^2 \|\Psi_{m,\gamma}^{-1} - \widehat{\Psi}_{m,\Gamma}^{-1}\|_{\mathrm{op}}^2\right\} \\
&\subset \bigcup_{\mathfrak{c}_{\varphi}^4 m^2 \leqslant \mathfrak{d}\frac{NT}{\log(NT)}} \left\{\|\widehat{\Psi}_{m,\Gamma} - \Psi_{m,\gamma}\|_{\mathrm{op}} > \frac{m}{2}\left(\frac{\log(NT)}{\mathfrak{d}NT}\right)^{\frac{1}{2}}\right\}.
\end{aligned}$$

Therefore, by Step 2, by Assumptions 3.2 and 3.4, and since $q > 2p$,

$$\begin{aligned}
\mathbb{P}(\widehat{\mathcal{M}}_{N,\Gamma} \not\subset \mathfrak{M}_{N,\gamma}) &\leqslant \sum_{\mathfrak{c}_{\varphi}^4 m^2 \leqslant \mathfrak{d}\frac{NT}{\log(NT)}} \left[2m\exp\left(-\frac{N}{128\mathfrak{L}(m)^2}\cdot\frac{m^2\log(NT)}{4\mathfrak{d}NT}\right)\right. \\
&\qquad\left.+\frac{8^q|\mathcal{I}_{N,\Gamma}|^q}{N^q}\cdot\frac{\mathfrak{d}^{q/2}(NT)^{q/2}}{m^q\log(NT)^{q/2}}\mathfrak{L}(m)^q\right] \\
&\leqslant \sum_{\mathfrak{c}_{\varphi}^4 m^2 \leqslant \mathfrak{d}\frac{NT}{\log(NT)}} [2m\exp\left(-\frac{\log(NT)}{512\mathfrak{c}_{\varphi}^4\mathfrak{d}T}\right) \\
&\qquad + (8\mathfrak{c}_{\varphi}^2)^q(\mathfrak{d}T)^{\frac{q}{2}}\frac{|\mathcal{I}_{N,\Gamma}|^q}{N^{q/2}\log(NT)^{q/2}}] \\
&\leqslant \sum_{\mathfrak{c}_{\varphi}^4 m^2 \leqslant \mathfrak{d}\frac{NT}{\log(NT)}} \left(\frac{2m}{N^{1+q/2}} + (8\mathfrak{c}_{\varphi}^2)^q(\mathfrak{d}T)^{\frac{q}{2}}\frac{\overline{\mathfrak{c}}_{3.2}}{N^p}\right) \leqslant \frac{\mathfrak{c}_2}{N^{p-1}},
\end{aligned}$$

where $\mathfrak{c}_2 > 0$ is a constant not depending on N.

□

The following theorem, which is the main result of this section, provides a risk bound on the adaptive estimator $\widehat{b}_{\widehat{m}}$.

Theorem 3.2 *Under Assumptions 3.2 (with $p \geqslant 5$), 3.4 and 3.5, there exists a constant $\mathfrak{c}_{3.2} > 0$, not depending on N, such that*

$$\mathbb{E}(\|\widehat{b}_{\widehat{m}} - b_0^I\|_N^2) \leqslant \mathfrak{c}_{3.2}\left(\min_{m\in\mathcal{M}_N}\left\{\min_{\tau\in\mathcal{S}_m}\|\tau - b_0^I\|_f^2 + \frac{m}{N}\left(1 + \frac{1}{N}\sum_{l\neq k}|R_{l,k}|\right)\right\} + \frac{1}{N}\right).$$

Proof By the definition of $\widehat{\Pi}_{\widehat{m}}$ (see (3.9)),

$$\|\widehat{b}_{\widehat{m}} - b_0^I\|_N^2 = \min_{\tau\in\mathcal{S}_{\widehat{m}}}\|\tau - b_0^I\|_N^2 + \|\widehat{b}_{\widehat{m}} - \widehat{\Pi}_{\widehat{m}}(b_0^I)\|_N^2.$$

Moreover, since $0 \in \mathcal{S}_{\widehat{m}}$,

$$\min_{\tau \in S_{\widehat{m}}} \|\tau - b_0^I\|_N^2 \leqslant \|b_0^I\|_N^2.$$

Then,

$$\|\widehat{b}_{\widehat{m}} - b_0^I\|_N^2 \leqslant (\|b_0^I\|_N^2 + \|\widehat{b}_{\widehat{m}} - \widehat{\Pi}_{\widehat{m}}(b_0^I)\|_N^2)(\mathbf{1}_{\Xi_N^c} + \mathbf{1}_{\Omega_N^c}) + \|\widehat{b}_{\widehat{m}} - b_0^I\|_N^2 \mathbf{1}_{\Xi_N \cap \Omega_N}$$

with

$$\Omega_N = \bigcap_{m \in \mathfrak{M}_N} \Omega_m.$$

The proof of Theorem 3.2 is dissected in four steps. The first step provides a preliminary bound on $\mathbb{E}(\|\widehat{b}_{\widehat{m}} - b_0^I\|_N^2)$, which is improved in Step 4 thanks to Propositions 3.6 and 3.8, and to the bound established in Steps 2 and 3 on

$$\rho_N(m) = \mathbb{E}\left(\left(\left[\sup_{\tau \in \mathcal{B}_{m,\widehat{m}}} |\nu_N(\tau)|\right]^2 - p(m,\widehat{m})\right)_+ \mathbf{1}_{\Xi_N \cap \Omega_N}\right) ; m \in \mathcal{M}_N$$

where, for every $m' \in \mathcal{M}_N$,

$$\mathcal{B}_{m,m'} = \{\tau \in \mathcal{S}_{m \vee m'} : \|\tau\|_f = 1\}$$

and

$$p(m,m') = \frac{\mathfrak{c}_{\text{cal}}}{8} \cdot \frac{m \vee m'}{N}\left(1 + \frac{1}{N}\sum_{i \neq k} |R_{i,k}|\right).$$

Step 1. On the one hand, since $\widehat{m} \in \widehat{\mathcal{M}}_N$, as established in the first step of the proof of Theorem 3.1,

$$\begin{aligned}
\|\widehat{b}_{\widehat{m}} - \widehat{\Pi}_{\widehat{m}}(b_0^I)\|_N^2 &= \widehat{\Delta}_{\widehat{m}}^* \widehat{\Psi}_{\widehat{m}}^{-1} \widehat{\Delta}_{\widehat{m}} = \|\widehat{\Psi}_{\widehat{m}}^{-\frac{1}{2}} \widehat{\Delta}_{\widehat{m}}\|_{2,\widehat{m}}^2 \\
&\leqslant \|\widehat{\Psi}_{\widehat{m}}^{-1}\|_{\text{op}} \sum_{j=1}^{\widehat{m}} \nu_N(\varphi_j)^2 \leqslant \left(\mathfrak{d}\frac{NT}{\log(NT)}\right)^{\frac{1}{2}} \sum_{j=1}^{N_T} \nu_N(\varphi_j)^2.
\end{aligned}$$

Then, by the second step of the proof of Theorem 3.1,

$$\begin{aligned}
\mathbb{E}(\|\widehat{b}_{\widehat{m}} - \widehat{\Pi}_{\widehat{m}}(b_0^I)\|_N^2 \mathbf{1}_{\Xi_N^c}) &\leqslant \left(\mathfrak{d}\frac{NT}{\log(NT)}\right)^{\frac{1}{2}} \mathbb{E}\left[\left(\sum_{j=1}^{N_T} \nu_N(\varphi_j)^2\right)^2\right]^{\frac{1}{2}} \mathbb{P}(\Xi_N^c)^{\frac{1}{2}} \\
&\leqslant \mathfrak{c}_1 \left(\frac{NT}{\log(NT)}\right)^{\frac{1}{2}} \frac{N_T^{1/2}\mathfrak{L}(N_T)}{NT}
\end{aligned}$$

$$\times \left(1 + \frac{1}{N}\sum_{i \neq k} |R_{i,k}|\right) \mathbb{P}(\Xi_N^c)^{\frac{1}{2}}$$

$$\leqslant \mathfrak{c}_2 N \left(1 + \frac{1}{N}\sum_{i \neq k} |R_{i,k}|\right) \mathbb{P}(\Xi_N^c)^{\frac{1}{2}}$$

and, in the same way,

$$\mathbb{E}(\|\widehat{b}_{\widehat{m}} - \widehat{\Pi}_{\widehat{m}}(b_0^I)\|_N^2 \mathbf{1}_{\Omega_N^c}) \leqslant \mathfrak{c}_3 N \left(1 + \frac{1}{N}\sum_{i \neq k} |R_{i,k}|\right) \mathbb{P}(\Omega_N^c)^{\frac{1}{2}},$$

where the $\mathfrak{c}_j$'s are positive constants not depending on N. On the other hand, for every $\tau, \overline{\tau} \in \mathfrak{S}_{N_T}$,

$$\gamma_N(\overline{\tau}) - \gamma_N(\tau) = \|\overline{\tau} - b_0^I\|_N^2 - \|\tau - b_0^I\|_N^2 - 2\nu_N(\overline{\tau} - \tau),$$

where

$$\mathfrak{S}_m = \mathcal{S}_1 \cup \cdots \cup \mathcal{S}_m; \forall m \in \{1, \dots, N_T\}.$$

Moreover, since

$$\widehat{m} = \arg\min_{m \in \widehat{\mathcal{M}}_N} \{\gamma_N(\widehat{b}_m) + \text{pen}(m)\},$$

for every $m \in \widehat{\mathcal{M}}_N$,

$$\gamma_N(\widehat{b}_{\widehat{m}}) + \text{pen}(\widehat{m}) \leqslant \gamma_N(\widehat{b}_m) + \text{pen}(m). \tag{3.14}$$

On the event $\Xi_N = \{\mathcal{M}_N \subset \widehat{\mathcal{M}}_N \subset \mathfrak{M}_N\}$, Inequality (3.14) remains true for every $m \in \mathcal{M}_N$. Then, on Ξ_N, for any $m \in \mathcal{M}_N$, since $\mathcal{S}_m + \mathcal{S}_{\widehat{m}} \subset \mathcal{S}_{m \vee \widehat{m}}$ under Assumption 3.4,

$$\begin{aligned}
\|\widehat{b}_{\widehat{m}} - b_0^I\|_N^2 &\leqslant \|\widehat{b}_m - b_0^I\|_N^2 + 2\nu_N(\widehat{b}_{\widehat{m}} - \widehat{b}_m) + \text{pen}(m) - \text{pen}(\widehat{m}) \\
&\leqslant \|\widehat{b}_m - b_0^I\|_N^2 + \frac{1}{8}\|\widehat{b}_{\widehat{m}} - \widehat{b}_m\|_f^2 + 8\left(\left[\sup_{\tau \in \mathcal{B}_{m,\widehat{m}}} |\nu_N(\tau)|\right]^2 - p(m, \widehat{m})\right)_+ \\
&\quad + \text{pen}(m) + 8p(m, \widehat{m}) - \text{pen}(\widehat{m}).
\end{aligned}$$

Since $\|\tau\|_f^2 \mathbf{1}_{\Omega_N} \leqslant 2\|\tau\|_N^2 \mathbf{1}_{\Omega_N}$ for every $\tau \in \mathfrak{S}_{\max(\mathfrak{M}_N)}$, and since $8p(m, \widehat{m}) \leqslant \text{pen}(m) + \text{pen}(\widehat{m})$,

$$\|\widehat{b}_{\widehat{m}} - b_0^I\|_N^2 \leqslant 3\|\widehat{b}_m - b_0^I\|_N^2 + 4\text{pen}(m)$$

$$+16\left(\left[\sup_{\tau\in\mathcal{B}_{m,\widehat{m}}}|\nu_N(\tau)|\right]^2-p(m,\widehat{m})\right)_+ \quad \text{on} \quad \Xi_N\cap\Omega_N.$$

So,

$$\mathbb{E}(\|\widehat{b}_{\widehat{m}}-b_0^I\|_N^2\mathbf{1}_{\Xi_N\cap\Omega_N})\leqslant\min_{m\in\mathcal{M}_N}\{3\mathbb{E}(\|\widehat{b}_m-b_0^I\|_N^2\mathbf{1}_{\Xi_N})+4\text{pen}(m)+16\rho_N(m)\}.$$

Therefore, to conclude this first step,

$$\begin{aligned}\mathbb{E}(\|\widehat{b}_{\widehat{m}}-b_0^I\|_N^2)&\leqslant\min_{m\in\mathcal{M}_N}\{3\mathbb{E}(\|\widehat{b}_m-b_0^I\|_N^2\mathbf{1}_{\Xi_N})+4\text{pen}(m)+16\rho_N(m)\}\\&\quad+\mathfrak{c}_4N\left(1+\frac{1}{N}\sum_{i\neq k}|R_{i,k}|\right)(\mathbb{P}(\Xi_N^c)^{\frac{1}{2}}+\mathbb{P}(\Omega_N^c)^{\frac{1}{2}}),\end{aligned}$$

where $\mathfrak{c}_4>0$ is a constant not depending on N.

Step 2. Consider $\tau\in\mathfrak{S}_{N_T}$ and the martingale $(M_N(\tau)_t)_{t\in[0,T]}$ defined by

$$M_N(\tau)_t=\sum_{i=1}^N\int_0^t\sigma(X_s^i)\tau(X_s^i)dW_s^i;\ \forall t\in[0,T].$$

Since $d\langle W^i,W^k\rangle_t=R_{i,k}dt$ for every $i,k\in\{1,\dots,N\}$,

$$\begin{aligned}\langle M_N(\tau)\rangle_T&=\sum_{i,k=1}^N R_{i,k}\int_0^T\sigma(X_s^i)\sigma(X_s^k)\tau(X_s^i)\tau(X_s^k)ds\\&=\int_0^T((\sigma\tau)(X_s^i))_i^*R((\sigma\tau)(X_s^i))_ids\leqslant\|R\|_{\text{op}}\int_0^T\|((\sigma\tau)(X_s^i))_i\|_{2,N}^2ds\\&\leqslant\|R\|_{\text{op}}\|\sigma\|_\infty^2\int_0^T\left(\sum_{i=1}^N\tau(X_s^i)^2\right)ds=NT\|R\|_{\text{op}}\|\sigma\|_\infty^2\|\tau\|_N^2.\end{aligned}$$

Then, by Assumption 3.5 and Bernstein's inequality for continuous local martingales (see Revuz and Yor [2], p. 153), for any $\varepsilon,\upsilon>0$,

$$\begin{aligned}\mathbb{P}(\nu_N(\tau)\geqslant\varepsilon,\|\tau\|_N^2\leqslant\upsilon^2)&\leqslant\mathbb{P}(M_N(\tau)_T^*\geqslant NT\varepsilon,\langle M_N(\tau)\rangle_T\leqslant NT\upsilon^2\|R\|_{\text{op}}\|\sigma\|_\infty^2)\\&\leqslant\exp\left(-\frac{NT\varepsilon^2}{2\upsilon^2\|\sigma\|_\infty^2\mathfrak{r}}\right)\quad\text{with}\quad\mathfrak{r}=1+\mathfrak{m}_{3.5}.\end{aligned}$$

Since this bound remains true by replacing τ by $-\tau$,

$$\begin{aligned}\mathbb{P}(|\nu_N(\tau)| \geqslant \varepsilon, \|\tau\|_N^2 \leqslant \upsilon^2) &= \mathbb{P}(\nu_N(\tau) \geqslant \varepsilon, \|\tau\|_N^2 \leqslant \upsilon^2)\\ &\quad + \mathbb{P}(\nu_N(-\tau) \geqslant \varepsilon, \|\tau\|_N^2 \leqslant \upsilon^2)\\ &\leqslant 2\exp\left(-\frac{NT\varepsilon^2}{2\upsilon^2\|\sigma\|_\infty^2 \mathfrak{t}}\right).\end{aligned}$$

Step 3. By using Step 2 and the same chaining technique as in the proof of Baraud et al. [7], Proposition 6.1, the purpose of this step is to find a suitable bound on

$$\mathbb{E}\left(\left(\left[\sup_{\tau\in\mathcal{B}_{m,m'}} |\nu_N(\tau)|\right]^2 - p(m,m')\right)_+ \mathbf{1}_{\Xi_N\cap\Omega_N}\right); m, m' \in \mathcal{M}_N.$$

Consider $\delta_0 \in (0, 1)$ and let $(\delta_n)_{n\in\mathbb{N}^*}$ be the sequence defined by

$$\delta_n = \delta_0 2^{-n}; \forall n \in \mathbb{N}^*.$$

Since $\mathcal{S}_{m\vee m'}$ is a vector subspace of $\mathbb{L}^2(\mathbb{R}, f(x)dx)$ of dimension $m \vee m'$, by Lorentz et al. [8], Chap. 15, Proposition 1.3, for any $n \in \mathbb{N}$, there exists $T_n \subset \mathcal{B}_{m,m'}$ such that $|T_n| \leqslant (3/\delta_n)^{m\vee m'}$ and, for any $\tau \in \mathcal{B}_{m,m'}$,

$$\exists \tau_n \in T_n : \|\tau - \tau_n\|_f \leqslant \delta_n. \tag{3.15}$$

In particular, note that

$$\tau = \tau_0 + \sum_{n=1}^{\infty}(\tau_n - \tau_{n-1}).$$

Then, for any sequence $(\Delta_n)_{n\in\mathbb{N}}$ of elements of $(0, \infty)$ such that $\Delta = \sum_{n\in\mathbb{N}} \Delta_n < \infty$,

$$\begin{aligned}&\left\{\left[\sup_{\tau\in\mathcal{B}_{m,m'}} |\nu_N(\tau)|\right]^2 > \Delta^2\right\}\\ &\quad\subset \left\{\exists(\tau_n)_{n\in\mathbb{N}} \in \prod_{n=0}^{\infty} T_n : |\nu_N(\tau_0)| + \sum_{n=1}^{\infty} |\nu_N(\tau_n - \tau_{n-1})| > \Delta\right\}\\ &\quad\subset \left\{\exists(\tau_n)_{n\in\mathbb{N}} \in \prod_{n=0}^{\infty} T_n : |\nu_N(\tau_0)| > \Delta_0 \text{ or } (\exists n \in \mathbb{N}^* : |\nu_N(\tau_n - \tau_{n-1})| > \Delta_n)\right\}\\ &\quad\subset \bigcup_{\tau_0\in T_0}\{|\nu_N(\tau_0)| > \Delta_0\} \cup \bigcup_{n=1}^{\infty} \bigcup_{(\tau_{n-1},\tau_n)\in\mathbb{T}_n} \{|\nu_N(\tau_n - \tau_{n-1})| > \Delta_n\}\end{aligned}$$

with $\mathbb{T}_n = T_{n-1} \times T_n$ for every $n \in \mathbb{N}^*$. Moreover, since

$$\|\varphi\|_N^2 \mathbf{1}_{\Omega_N} \leqslant \frac{3}{2}\|\varphi\|_f^2 \mathbf{1}_{\Omega_N}; \ \forall \varphi \in \mathfrak{S}_{\max(\mathfrak{M}_N)},$$

for every $(\tau_n)_{n\in\mathbb{N}} \in \prod_{n\in\mathbb{N}} T_n$ satisfying (3.15) for a given $\tau \in \mathcal{B}_{m,m'}$, on Ω_N,

$$\|\tau_0\|_N^2 \leqslant \frac{3}{2}\|\tau_0\|_f^2 \leqslant \frac{3}{2}$$

and, for every $n \in \mathbb{N}^*$,

$$\|\tau_n - \tau_{n-1}\|_N^2 \leqslant \frac{3}{2}\|\tau_n - \tau_{n-1}\|_f^2 \leqslant \frac{3}{2}(2\delta_{n-1}^2 + 2\delta_n^2) = \frac{15}{4}\delta_{n-1}^2.$$

So, by Step 2,

$$\begin{aligned}\mathbb{P}&\left(\left\{\left[\sup_{\tau\in\mathcal{B}_{m,m'}} |\nu_N(\tau)|\right]^2 > \Delta^2\right\} \cap \Xi_N \cap \Omega_N\right)\\ &\leqslant 2\sum_{\tau_0\in T_0} \exp\left(-\frac{NT\Delta_0^2}{3\|\tau_0\|_f^2\|\sigma\|_\infty^2\mathfrak{r}}\right)\\ &\qquad + 2\sum_{n=1}^{\infty}\sum_{(\tau_{n-1},\tau_n)\in\mathbb{T}_n} \exp\left(-\frac{NT\Delta_n^2}{3\|\tau_n-\tau_{n-1}\|_f^2\|\sigma\|_\infty^2\mathfrak{r}}\right)\\ &\leqslant 2\exp\left(h_0 - \frac{NT\Delta_0^2}{3\|\sigma\|_\infty^2\mathfrak{r}}\right)\\ &\qquad + 2\sum_{n=1}^{\infty}\exp\left(h_{n-1} + h_n - \frac{NT\Delta_n^2}{15/2\delta_{n-1}^2\|\sigma\|_\infty^2\mathfrak{r}}\right)\end{aligned} \tag{3.16}$$

with $h_n = \log(|T_n|)$ for every $n \in \mathbb{N}$. Now, let us take Δ_0 such that

$$h_0 - \frac{NT\Delta_0^2}{3\|\sigma\|_\infty^2\mathfrak{r}} = -(m \vee m' + x) \quad \text{with} \quad x > 0,$$

leading to

$$\Delta_0 = \left(\frac{3\|\sigma\|_\infty^2\mathfrak{r}}{NT}(m \vee m' + x + h_0)\right)^{\frac{1}{2}},$$

and for every $n \in \mathbb{N}^*$, let us take Δ_n such that

$$h_{n-1} + h_n - \frac{NT\Delta_n^2}{15/2\delta_{n-1}^2\|\sigma\|_\infty^2\mathfrak{r}} = -(m \vee m' + x + n),$$

leading to

$$\Delta_n = \left(\frac{15/2\delta_{n-1}^2\|\sigma\|_\infty^2\mathfrak{r}}{NT}(m \vee m' + x + n + h_{n-1} + h_n)\right)^{\frac{1}{2}}.$$

For this appropriate sequence $(\Delta_n)_{n\in\mathbb{N}}$,

$$\begin{aligned}\mathbb{P}\left(\left\{\left[\sup_{\tau\in\mathcal{B}_{m,m'}}|\nu_N(\tau)|\right]^2 > \Delta^2\right\}\cap \Xi_N \cap \Omega_N\right) &\leqslant 2e^{-x}e^{-(m\vee m')}\left(1+\sum_{n=1}^{\infty}e^{-n}\right)\\ &\leqslant 3.2e^{-x}e^{-(m\vee m')}\end{aligned}$$

by Inequality (3.16), and

$$\begin{aligned}\Delta^2 &\leqslant \frac{3\|\sigma\|_\infty^2\mathfrak{r}}{NT}((m \vee m' + x)^{\frac{1}{2}} + h_0^{\frac{1}{2}}\\ &\qquad + \sqrt{\frac{5}{2}}\sum_{n=1}^{\infty}\delta_{n-1}((m \vee m' + x)^{\frac{1}{2}} + (n + h_{n-1} + h_n)^{\frac{1}{2}}))^2\\ &\leqslant \frac{3\|\sigma\|_\infty^2\mathfrak{r}}{NT}(m \vee m' + x)(1 + h_0^{\frac{1}{2}}\\ &\qquad + \sqrt{\frac{5}{2}}\sum_{n=1}^{\infty}\delta_{n-1}(1 + (n + h_{n-1} + h_n)^{\frac{1}{2}}))^2\\ &\leqslant \frac{3\|\sigma\|_\infty^2\mathfrak{r}}{NT}(m \vee m' + x)(\delta + \overline{\delta})\end{aligned}$$

with

$$\delta = 2\left(1 + \sqrt{\frac{5}{2}}\sum_{n=1}^{\infty}\delta_{n-1}\right)^2$$

and

$$\overline{\delta} = 2\left[h_0^{\frac{1}{2}} + \sqrt{\frac{5}{2}}\sum_{n=1}^{\infty}\delta_{n-1}\left(n + N_T\left(2\log\left(\frac{3}{\delta_0}\right) + (2n-1)\log(2)\right)\right)^{\frac{1}{2}}\right]^2,$$

because

$$\begin{aligned}h_{n-1} + h_n &\leqslant (m \vee m')\left(\log\left(\frac{3}{\delta_{n-1}}\right) + \log\left(\frac{3}{\delta_n}\right)\right)\\ &\leqslant N_T\left(2\log\left(\frac{3}{\delta_0}\right) + (2n-1)\log(2)\right).\end{aligned}$$

Then,

$$\mathbb{P}\left(\left\{\left[\sup_{\tau\in\mathcal{B}_{m,m'}}|\nu_N(\tau)|\right]^2-\frac{8\rho}{\mathfrak{c}_{\text{cal}}TR_N}p(m,m')>\frac{\rho}{NT}x\right\}\cap\Xi_N\cap\Omega_N\right)$$
$$\leqslant 3.2e^{-x}e^{-(m\vee m')}$$

with

$$\rho=3\|\sigma\|_\infty^2\mathfrak{r}(\delta+\overline{\delta})\quad\text{and}\quad R_N=1+\frac{1}{N}\sum_{i\neq k}|R_{i,k}|.$$

So, by taking $\mathfrak{c}_{\text{cal}}\geqslant 8\rho T^{-1}>8\rho(TR_N)^{-1}$ and $y=\rho x(NT)^{-1}$,

$$p(y)=\mathbb{P}\left(\left\{\left[\sup_{\tau\in\mathcal{B}_{m,m'}}|\nu_N(\tau)|\right]^2-p(m,m')>y\right\}\cap\Xi_N\cap\Omega_N\right)$$
$$\leqslant 3.2e^{-\frac{NTy}{\rho}}e^{-(m\vee m')}.$$

Therefore,

$$\mathbb{E}\left(\left(\left[\sup_{\tau\in\mathcal{B}_{m,m'}}|\nu_N(\tau)|\right]^2-p(m,m')\right)_+\mathbf{1}_{\Xi_N\cap\Omega_N}\right)=\int_0^\infty p(y)dy$$
$$\leqslant 3.2\rho\frac{e^{-(m\vee m')}}{NT}.$$

Step 4 (conclusion). By Step 3, there exists a constant $\mathfrak{c}_5>0$, not depending on N, such that for every $m\in\mathcal{M}_N$,

$$\rho_N(m)\leqslant\frac{3.2\rho}{NT}\sum_{m'\in\mathcal{M}_N}e^{-(m\vee m')}\leqslant\frac{\mathfrak{c}_5}{N}.$$

So, under Assumptions 3.4 and 3.5, by Step 1,

$$\mathbb{E}(\|\widehat{b}_{\widehat{m}}-b_0^I\|_N^2)\leqslant\min_{m\in\mathcal{M}_N}\left\{3\mathbb{E}(\|\widehat{b}_m-b_0^I\|_N^2\mathbf{1}_{\Xi_N})+4\text{pen}(m)+\frac{16\mathfrak{c}_5}{N}\right\}$$
$$+\mathfrak{c}_4N\left(1+\frac{1}{N}\sum_{i\neq k}|R_{i,k}|\right)(\mathbb{P}(\Xi_N^c)^{\frac{1}{2}}+\mathbb{P}(\Omega_N^c)^{\frac{1}{2}}).$$

Under the additional Assumption 3.2 with $p\geqslant 5$, by Propositions 3.6 (because $\widetilde{\mathfrak{d}}\leqslant\overline{\mathfrak{c}}_\Lambda$) and 3.8, there exists a constant $\mathfrak{c}_6>0$, not depending on N, such that

$$\mathbb{E}(\|\widehat{b}_{\widehat{m}} - b_0^I\|_N^2) \leqslant \mathfrak{c}_6 \left(\min_{m \in \mathcal{M}_N} \{\mathbb{E}(\|\widehat{b}_m - b_0^I\|_N^2 \mathbf{1}_{\Xi_N}) + \text{pen}(m)\} + \frac{1}{N} \right),$$

and then Theorem 3.1 allows to conclude.

□

Let us conclude this section with some remarks about Theorem 3.2:

- Note that Theorem 3.2 provides a risk bound on the adaptive estimator $\widehat{b}_{\widehat{m}}$ of same order as the minimal risk bound on $\widetilde{b}_m$ (see Inequality (3.10)) for m taken in $\mathcal{M}_N$.
- Of course the model selection criterion (3.13) works when $W^1, \ldots, W^N$ are independent, but in this special case, the criterion may be improved by taking a larger $\widehat{\mathcal{M}}_N$:

  $$\widehat{m}^* = \arg \min_{m \in \widehat{\mathcal{M}}_N^*} \{-\|\widehat{b}_m\|_N^2 + \text{pen}(m)\},$$

  where

  $$\widehat{\mathcal{M}}_N^* = \left\{ m \in \{1, \ldots, N_T\} : \mathfrak{c}_\varphi^2 m(\|\widehat{\Psi}_m^{-1}\|_{\text{op}} \vee 1)^2 \leqslant \mathfrak{d}^* \frac{NT}{\log(NT)} \right\}$$

  and $\mathfrak{d}^* \in (0, \bar{\mathfrak{c}}_\Lambda/4)$. Under Assumption 3.4, F. Comte and V. Genon-Catalot have established in [9] (see Theorem 3.1) that

  $$\mathbb{E}(\|\widehat{b}_{\widehat{m}^*} - b_0^I\|_N^2) \leqslant \mathfrak{c}_{3.2}^* \left(\min_{m \in \mathcal{M}_N^*} \left\{ \min_{\tau \in S_m} \|\tau - b_0^I\|_f^2 + \frac{m}{N} \right\} + \frac{1}{N} \right),$$

  where $\mathfrak{c}_{3.2}^* > 0$ is a constant not depending on N, and

  $$\mathcal{M}_N^* = \left\{ m \in \{1, \ldots, N_T\} : \mathfrak{c}_\varphi^2 m(\|\Psi_m^{-1}\|_{\text{op}} \vee 1)^2 \leqslant \bar{\mathfrak{d}}^* \frac{NT}{\log(NT)} \right\} \quad \text{with} \quad \bar{\mathfrak{d}}^* = \frac{\mathfrak{d}^*}{4}.$$

  This sharper result is established in the more general framework of diffusion processes with jumps in Chap. 4, Sect. 4.1.

References

1. Menozzi, S., Pesce, A. and Zhang, X. (2021). Density and Gradient Estimates for Non Degenerate Brownian SDEs with Unbounded Measurable Drift. *Journal of Differential Equations* **272**, 339–369.
2. Revuz, D. and Yor, M. (1999). *Continuous Martingales and Brownian Motion*. Springer, Berlin-Heidelberg.
3. Tropp, J.A. (2012). User-Friendly Tail Bounds for Sums of Random Matrices. *Foundations of Computational Mathematics* **12**, 389–434.

4. Cohen, A., Davenport, M.A. and Leviatan, D. (2013). On the Stability and Accuracy of Least Squares Approximation. *Foundations of Computational Mathematics* **13**, 819–834.
5. Stewart, G.W. and Sun, J.-G. (1990). *Matrix Perturbation Theory*. Academic Press, USA.
6. DeVore, R.A. and Lorentz, G.G. (1993). *Constructive Approximation*. Springer, Berlin-Heidelberg.
7. Baraud, Y., Comte, F. and Viennet, G. (2001). Model Selection for (Auto-)Regression with Dependent Data. *ESAIM:PS* **5**, 33–49.
8. Lorentz, G., von Golitschek, M. and Makokov, Y. (1996). *Constructive Approximation: Advanced Problems*. Springer, Berlin-Heidelberg.
9. Comte, F. and Genon-Catalot, V. (2020). Nonparametric Drift Estimation for I.I.D. Paths of Stochastic Differential Equations. *The Annals of Statistics* **48**(6), 3336–3365.

Chapter 4
Going Further with the Projection Least Squares Method: Diffusions with Jumps and Fractional Diffusions

The risk bound in Theorem 3.1 can be extended to stochastic differential equations driven by a jump process or by a fractional Brownian motion, because its proof only relies on the zero mean and on the control of the fourth-order moment of Itô's integral with respect to W, and because the bounds in Sect. 3.3.1 (resp. Sect. 3.4) on $\mathbb{P}(\Omega_{m,\Gamma}^c)$ (resp. $\mathbb{P}(\Lambda_{m,\Gamma}^c)$) and Lemma 3.4 have been established for copies of any piecewise continuous stochastic process. So, this chapter deals with the extensions of the projection least squares method introduced in Chap. 3 to diffusions with jumps and to fractional diffusions, but for independent copies only. On the one hand, Theorems 3.1 and 3.2, which respectively provide a risk bound on $\widehat{b}_m$ for m fixed and a risk bound on the adaptive estimator $\widehat{b}_{\widehat{m}}$, are extended to the projection least squares estimator of the drift function for stochastic differential equations driven by a Lévy process in Sect. 4.1. On the other hand, to generalize the (projection) least squares method to fractional diffusions is a more complicated challenge, because the natural extension of Itô's integral to fractional Brownian motion, centered with a sharp control of its fourth-order moment, is the Skorokhod integral, which is difficult to compute in practice. This difficulty is deeply discussed in Sect. 4.2, which deals with a risk bound on a copies-based parametric least squares estimator for fractional diffusions, with a risk bound on the Skorokhod integral-based extension of the projection least squares estimator introduced in Chap. 3, and finally with a projection least squares estimator of the drift function for some non-autonomous fractional diffusions.

4.1 The Projection Least Squares Estimator of the Drift Function for Diffusions with Jumps

Let $X = (X_t)_{t\in\mathbb{R}_+}$ be the diffusion process with jumps defined by

$$X_t = x_0 + \int_0^t b_0(X_s)ds + \int_0^t \sigma(X_s)dW_s + \int_0^t \gamma(X_s)d\mathfrak{Z}_s;\ t \in \mathbb{R}_+ \qquad (4.1)$$

N. Marie, *From Nonparametric Regression to Statistical Inference for Non-Ergodic Diffusion Processes*, Frontiers in Probability and the Statistical Sciences,
https://doi.org/10.1007/978-3-031-95638-6_4

where $x_0 \in \mathbb{R}$, $W = (W_t)_{t\in\mathbb{R}_+}$ is a Brownian motion, $\mathfrak{Z} = (\mathfrak{Z}_t)_{t\in\mathbb{R}_+}$ is a compensated compound Poisson process independent of W, $b_0 \in C^1(\mathbb{R})$ with b_0' bounded, and $\sigma, \gamma : \mathbb{R} \to \mathbb{R}$ are bounded Lipschitz continuous functions such that

$$\left(\inf_{x\in\mathbb{R}} \sigma(x)^2\right) \wedge \left(\inf_{x\in\mathbb{R}} \gamma(x)^2\right) > 0.$$

Under these conditions on b_0, σ and γ, Eq. (4.1) has a unique (strong) solution.

In this section, except those depending on the empirical process ν_N, all the operators (as γ_N or $\widehat{\Psi}_m$) and sets (as Λ_m) introduced in Chap. 3 remain defined in the same way by replacing the copies of the solution of Eq. (1.1) by $X^1, \dots, X^N$, where

$$X^i = \mathcal{I}(x_0, W^i, \mathfrak{Z}^i); \forall i \in \{1, \dots, N\},$$

$(W^1, \mathfrak{Z}^1), \dots, (W^N, \mathfrak{Z}^N)$ are N independent copies of $(W, \mathfrak{Z})$, and $\mathcal{I}$ is the solution map for Eq. (4.1). The projection least squares estimator of b_0 is still defined as a minimizer $\widehat{b}_m$ in $\mathcal{S}_m$ of the objective function γ_N, and the associated truncated estimator $\widetilde{b}_m$ by

$$\widetilde{b}_m(x) = \widehat{b}_m(x)\mathbf{1}_{\Lambda_m}; x \in I.$$

First of all, Sect. 4.1.1 is a brief reminder on the stochastic integral with respect to the compound Poisson process. In Sect. 4.1.2, a risk bound on $\widetilde{b}_m$ is established for m fixed. Section 4.1.3 deals with a model selection method and a risk bound on the associated adaptive estimator. Finally, some basic numerical experiments on the adaptive projection least squares estimator are presented in Sect. 4.1.4.

4.1.1 A Brief Reminder on the Stochastic Integral with Respect to the Compound Poisson Process

Let $Z = (Z_t)_{t\in[0,T]}$ be a compound Poisson process:

$$Z_t = \sum_{n=1}^{\nu_t} \zeta_n; \forall t \in [0, T],$$

where $\nu = (\nu_t)_{t\in[0,T]}$ is a (usual) Poisson process of intensity $\lambda > 0$, independent of the ζ_n's which are i.i.d. random variables of (common) probability distribution π. Let us also consider the centered martingale $\mathfrak{Z} = (\mathfrak{Z}_t)_{t\in[0,T]}$ defined by

$$\mathfrak{Z}_t = Z_t - \mathfrak{c}_\zeta \lambda t; \forall t \in [0, T],$$

where $\mathfrak{c}_\zeta$ is the (common) expectation of the ζ_n's. Its quadratic variation $[\![\mathfrak{Z}]\!] = ([\![\mathfrak{Z}]\!]_t)_{t\in[0,T]}$ satisfies

$$[\![\mathfrak{Z}]\!]_t = \sum_{i=1}^{\nu_t} \zeta_i^2;\ \forall t \in [0, T].$$

The stochastic integral on $[0, T]$ of a process $H = (H_t)_{t\in[0,T]}$ with respect to Z is defined by

$$\int_0^T H_s dZ_s = \sum_{n=1}^{\nu_T} H_{\tau_n}\zeta_n,$$

where the τ_n's are the jump times of the Poisson process ν. As Itô's integral, the stochastic integral with respect to $\mathfrak{Z}$ satisfies the two following basic properties: for every $H \in \mathbb{H}^2$ having left limits,

(A) $\int_0^{\cdot} H_s d\mathfrak{Z}_s = \left(\int_0^T H_s \mathbf{1}_{[0,t]}(s) d\mathfrak{Z}_s\right)_{t\in[0,T]}$ is a martingale,

(B) and $\mathbb{E}\left[\left(\int_0^T H_s d\mathfrak{Z}_s\right)^2\right] = \mathfrak{c}_{\zeta^2}\lambda \int_0^T \mathbb{E}(H_s^2)ds$ (isometry type property).

As for Itô's integral again, by the Burkholder-Davis-Gundy inequality,

(C) For every $p \geqslant 1$, there exists a constant $\mathfrak{b}_p > 0$ such that, for every $H \in \mathbb{H}^2$ having left limits,

$$\mathbb{E}\left(\sup_{t\in[0,T]}\left|\int_0^t H_s d\mathfrak{Z}_s\right|^p\right) \leqslant \mathfrak{b}_p \mathbb{E}\left(\left[\!\!\left[\int_0^{\cdot} H_s d\mathfrak{Z}_s\right]\!\!\right]_T^{\frac{p}{2}}\right)$$

and

$$\left[\!\!\left[\int_0^{\cdot} H_s d\mathfrak{Z}_s\right]\!\!\right]_T = \int_0^T H_s^2 d[\![\mathfrak{Z}]\!]_s.$$

Finally, for $p = 4$, the following proposition deals with a suitable bound on the left-hand side of Inequality (C).

Proposition 4.1 *For every $H \in \mathbb{H}^2$ having left limits,*

$$\mathbb{E}\left[\left(\int_0^T H_s^2 d[\![\mathfrak{Z}]\!]_s\right)^2\right] \leqslant 2(\mathfrak{c}_{\zeta^4} + \mathfrak{c}_{\zeta^2}^2\lambda T)\lambda \int_0^T \mathbb{E}(H_s^4)ds.$$

Proof Consider $H \in \mathbb{H}^2$ having left limits. By the isometry type property (B), and since $[\![\mathfrak{Z}]\!]$ is (also) a compound Poisson process of intensity λ,

$$\mathbb{E}\left[\left(\int_0^T H_s^2 d[\![3]\!]_s\right)^2\right] \leqslant 2\mathbb{E}\left[\left(\int_0^T H_s^2(d[\![3]\!]_s - \mathfrak{c}_{\zeta^2}\lambda ds)\right)^2\right]$$
$$+ 2\mathbb{E}\left[\left(\mathfrak{c}_{\zeta^2}\lambda\int_0^T H_s^2 ds\right)^2\right]$$
$$= 2\mathfrak{c}_{\zeta^4}\lambda\int_0^T \mathbb{E}(H_s^4)ds + 2\mathfrak{c}_{\zeta^2}^2\lambda^2\mathbb{E}\left[\left(\int_0^T H_s^2 ds\right)^2\right]$$
$$\leqslant 2(\mathfrak{c}_{\zeta^4} + \mathfrak{c}_{\zeta^2}^2\lambda T)\lambda\int_0^T \mathbb{E}(H_s^4)ds.$$

□

For details on the stochastic integral with respect to the compound Poisson process, the reader may refer to Applebaum [1] and Privault [2], Chap. 20.

4.1.2 Nonadaptive Risk Bound

By using the bounds on $\mathbb{P}(\Omega_{m,\Gamma}^c)$ and $\mathbb{P}(\Lambda_{m,\Gamma}^c)$ established in Sects. 3.3.1 and 3.4 respectively, and by using Lemma 3.4, the purpose of this section is to prove a suitable risk bound on $\widetilde{b}_m$ for m fixed. First, let us check that the density function f remains well-defined, and let us provide an appropriate empirical process.

In the sequel, σ satisfies (1.3), b_0 belongs to the Kato class

$$\mathbb{K}_2 = \left\{\varphi : \lim_{\delta\to 0}\sup_{x\in\mathbb{R}}\int_0^\delta\int_{-\infty}^{\infty}|\varphi(x+y)+\varphi(x-y)|s^{\frac{1}{2}}(|y|+s^{\frac{1}{2}})^{-3}dyds = 0\right\},$$

the Lévy measure $\pi_\lambda = \lambda\pi$ has a density θ with respect to Lebesgue's measure, there exists $\alpha\in(0,2)$ such that $z\in\mathbb{R}\mapsto\theta(z)|z|^{1+\alpha}$ is bounded, and if $\alpha = 1$, then

$$\int_{r<|z|<r+\varepsilon} z\theta(z)dz = 0;\ \forall r,\varepsilon > 0.$$

By Chen et al. [3], Theorem 1.1 and the remark p. 126, l. 5–7, in Amorino and Gloter [4], for every $t\in(0,T]$, the probability distribution of X_t has a density f_t with respect to Lebesgue's measure such that, for every $x\in\mathbb{R}$,

$$f_t(x) \leqslant \overline{\mathfrak{c}}_{0.5,T}\left(t^{-\frac{1}{2}}\exp\left(-\overline{\mathfrak{m}}_{0.5,T}\frac{(x-x_0)^2}{t}\right) + \frac{t}{(t^{1/2}+|x-x_0|)^{1+\alpha}}\right) \tag{4.2}$$

where $\overline{\mathfrak{c}}_{0.5,T}$ and $\overline{\mathfrak{m}}_{0.5,T}$ are positive constants depending on T but not on t and x. So, $t\mapsto f_t(x)$ $(x\in\mathbb{R})$ belongs to $\mathbb{L}^1([0,T])$, which legitimates to consider again the density function f defined by

$$f(x) = \frac{1}{T}\int_0^T f_s(x)ds;\ \forall x \in \mathbb{R}.$$

Let us establish three consequences of Inequality (4.2):

- The density function f is bounded, and then $\mathcal{S}_m \subset \mathbb{L}^2(\mathbb{R}, f(x)dx)$ because the φ_j's belong to $\mathbb{L}^2(\mathbb{R}, dx)$. Indeed, since

$$-\frac{1}{2} < 1 - \frac{1+\alpha}{2} < \frac{1}{2},$$

for every $x \in \mathbb{R}$,

$$0 \leqslant f(x) \leqslant \frac{\overline{\mathfrak{c}}_{0.5,T}}{T}\left(\int_0^T s^{-\frac{1}{2}}ds + \int_0^T s^{1-\frac{1+\alpha}{2}}ds\right)$$
$$= \frac{\overline{\mathfrak{c}}_{0.5,T}}{T}\left(2T^{\frac{1}{2}} + \frac{T^{2-(1+\alpha)/2}}{2-(1+\alpha)/2}\right) < \infty.$$

- If b_0 is bounded, then $b_0 \in \mathbb{K}_2$ and

$$|b_0|^{\kappa} \in \mathbb{L}^2(\mathbb{R}, f(x)dx);\ \forall \kappa > 0.$$

First, since f is a density function, for every $\kappa > 0$,

$$\int_{-\infty}^{\infty}(|b_0(x)|^{\kappa})^2 f(x)dx \leqslant \|b_0\|_{\infty}^{2\kappa} < \infty.$$

Now, for any $y \in \mathbb{R}^*$ and $\delta > 0$,

$$\int_0^{\delta} s^{\frac{1}{2}}(|y| + s^{\frac{1}{2}})^{-3}ds \leqslant \frac{1}{|y|^3}\int_0^{\delta} s^{\frac{1}{2}}ds = \frac{2}{3}\cdot\frac{\delta^{3/2}}{|y|^3},$$

and if $0 < y^2 \leqslant \delta$, then

$$\int_0^{\delta} s^{\frac{1}{2}}(|y| + s^{\frac{1}{2}})^{-3}ds = \int_0^{y^2}\frac{s^{1/2}}{(|y| + s^{1/2})^3}ds + \int_{y^2}^{\delta}\frac{s^{1/2}}{(|y| + s^{1/2})^3}ds$$
$$\leqslant \frac{2}{3} + \int_1^{\delta/y^2}\frac{s^{1/2}}{(1+s^{1/2})^3}ds \leqslant \frac{2}{3} + \mathfrak{c}_1\left(\frac{\delta}{y^2}\right)^{\frac{1}{4}}$$

with

$$\mathfrak{c}_1 = \int_1^{\infty}\frac{s^{1/4}}{(1+s^{1/2})^3}ds < \infty.$$

So, for every $\delta \in (0, 1]$,

$$\sup_{x\in\mathbb{R}}\int_0^\delta\int_{-\infty}^{\infty}|b_0(x+y)+b_0(x-y)|s^{\frac{1}{2}}(|y|+s^{\frac{1}{2}})^{-3}dyds$$

$$\leqslant 2\|b_0\|_\infty\left[\int_{0<y^2\leqslant\delta}\left[\frac{2}{3}+\mathfrak{c}_1\left(\frac{\delta}{y^2}\right)^{\frac{1}{4}}\right]dy+\frac{2}{3}\int_{\delta<y^2}\frac{\delta^{3/2}}{|y|^3}dy\right]\leqslant \mathfrak{c}_2\delta^{\frac{1}{2}}$$

with

$$\mathfrak{c}_2=4\|b_0\|_\infty\left(\frac{2}{3}+\mathfrak{c}_1\int_0^1\frac{dy}{y^{1/2}}+\frac{2}{3}\int_1^\infty\frac{dy}{y^3}\right)<\infty.$$

Therefore, $b_0\in\mathbb{K}_2$.

- Assume that there exist $\mathfrak{c}>0$, $\kappa>0$ and $\varepsilon\in(0,\alpha)$ as close as possible to 0, such that

$$|b_0(x)|\leqslant\mathfrak{c}(1+|x|)^{\frac{\alpha-\varepsilon}{\kappa}};\forall x\in\mathbb{R}.$$

So,

$$|b_0(x)|^\kappa f_t(x)\underset{x\to\pm\infty,t\to 0^+}{=}\mathrm{O}\left(\frac{t^{-1/2}}{|x-x_0|^{1+\varepsilon}}\right)\quad\text{by Inequality (4.2),}$$

and then

$$\int_{-\infty}^{\infty}|b_0(x)|^\kappa f(x)dx<\infty.$$

Now, consider the empirical process ν_N defined by

$$\nu_N(\tau):=\frac{1}{NT}\sum_{i=1}^N\int_0^T\tau(X_s^i)(\sigma(X_s^i)dW_s^i+\gamma(X_s^i)d\mathfrak{Z}_s^i).$$

Note that Equality (3.8) remains true:

$$[\widehat{Z}_m]_j=\langle b_0,\varphi_j\rangle_N+\nu_N(\varphi_j);\forall j\in\{1,\dots,m\}.$$

The following theorem, which is the main result of this section, provides a risk bound on $\widetilde{b}_m$.

Theorem 4.1 *If the matrix Ψ_m is invertible, then there exists a constant $\mathfrak{c}_{4.1,1}>0$, not depending on m and N, such that*

$$\mathbb{E}(\|\widetilde{b}_m-b_0^I\|_N^2)\leqslant\min_{\tau\in\mathcal{S}_m}\|\tau-b_0^I\|_f^2+\mathfrak{c}_{4.1,1}m\left(\frac{1}{N}+\mathbb{P}(\Omega_m^c)^{\frac{1}{2}}+\mathbb{P}(\Lambda_m^c)^{\frac{1}{2}}\right).$$

Under Assumption 3.1 with $p\geqslant 2$, there exists a constant $\mathfrak{c}_{4.1,2}>0$, not depending on m and N, such that

$$\mathbb{E}(\|\widetilde{b}_m-b_0^I\|_N^2)\leqslant\min_{\tau\in\mathcal{S}_m}\|\tau-b_0^I\|_f^2+\mathfrak{c}_{4.1,2}\frac{m}{N}.$$

Proof Since the definition of the empirical projection $\widehat{\Pi}_m$ remains the same as in Chap. 3, as in the proof of Theorem 3.1,

$$\mathbb{E}(\|\widetilde{b}_m - b_0^I\|_N^2) \leqslant \min_{\tau \in \mathcal{S}_m} \mathbb{E}(\|\tau - b_0^I\|_N^2) + \mathbb{E}(\|\widehat{b}_m - \widehat{\Pi}_m(b_0^I)\|_N^2 \mathbf{1}_{\Lambda_m}) + \mathfrak{c}_1 \mathbb{P}(\Lambda_m^c)^{\frac{1}{2}}$$

with

$$\mathfrak{c}_1 = \left(\int_I b_0(x)^4 f(x)dx\right)^{\frac{1}{2}}.$$

Moreover, by (3.8), and since Lemma 3.4 has been established for copies of any piecewise continuous stochastic process, as established in the proof of Theorem 3.1 (see the first step),

$$\begin{aligned}\mathbb{E}(\|\widehat{b}_m - \widehat{\Pi}_m(b_0^I)\|_N^2 \mathbf{1}_{\Lambda_m}) \leqslant\; & \frac{2}{NT}\text{trace}(\Psi_m^{-1}\Phi_{m,\sigma}) \\ & + \mathfrak{c}_\Lambda \frac{NT}{\log(NT)\mathfrak{L}(m)} \mathbb{E}\left[\left(\sum_{j=1}^m \nu_N(\varphi_j)^2\right)^2\right]^{\frac{1}{2}} \mathbb{P}(\Omega_m^c)^{\frac{1}{2}}\end{aligned}$$

with

$$\Phi_{m,\sigma} = NT\mathbb{E}(\widehat{\Delta}_m \widehat{\Delta}_m^*) \quad \text{and} \quad \widehat{\Delta}_m = (\nu_N(\varphi_j))_{j \in \{1,\dots,m\}}.$$

Since the empirical process is not the same as in Chap. 3, suitable bounds on

$$\mathbb{E}\left[\left(\sum_{j=1}^m \nu_N(\varphi_j)^2\right)^2\right] \quad \text{and} \quad \text{trace}(\Psi_m^{-1}\Phi_{m,\sigma})$$

need to be established (see Steps 1 and 2) in order to conclude (see Step 3).

Step 1. By Jensen's inequality,

$$\begin{aligned}\mathbb{E}\left[\left(\sum_{j=1}^m \nu_N(\varphi_j)^2\right)^2\right] &\leqslant m\sum_{j=1}^m \mathbb{E}(\nu_N(\varphi_j)^4) \\ &\leqslant \frac{8m}{N^4T^4}\sum_{j=1}^m (\mathbb{E}(\delta_j^4) + \mathbb{E}(\rho_j^4))\end{aligned}$$

where, for every $j \in \{1,\dots,m\}$,

$$\delta_j = \sum_{i=1}^N \int_0^T \varphi_j(X_s^i)\sigma(X_s^i)dW_s^i \quad \text{and} \quad \rho_j = \sum_{i=1}^N \int_0^T \varphi_j(X_s^i)\gamma(X_s^i)d\mathfrak{Z}_s^i.$$

On the one hand, as already established in the proof of Theorem 3.1 (see Step 2),

$$\sum_{j=1}^{m}\mathbb{E}(\delta_j^4) \leqslant \mathfrak{b}_4 N^2 T^2 \mathfrak{L}(m)^2 \int_{-\infty}^{\infty}\sigma(x)^4 f(x)dx.$$

On the other hand, by Burkholder-Davis-Gundy's inequality (see Sect. 4.1.1), Jensen's inequality and Proposition 4.1,

$$\begin{aligned}\sum_{j=1}^{m}\mathbb{E}(\rho_j^4) &\leqslant \mathfrak{b}_4 N^2 \sum_{j=1}^{m}\mathbb{E}\left[\left(\int_0^T \varphi_j(X_s)^2\gamma(X_s)^2 d[\![\mathfrak{Z}]\!]_s\right)^2\right]\\ &\leqslant 2\mathfrak{b}_4 N^2(\mathfrak{c}_{\zeta^4}+\mathfrak{c}_{\zeta^2}^2\lambda T)\lambda\\ &\quad\times\sum_{j=1}^{m}\int_0^T \mathbb{E}(\varphi_j(X_s)^4\gamma(X_s)^4)ds\\ &\leqslant 2\mathfrak{b}_4 N^2(\mathfrak{c}_{\zeta^4}+\mathfrak{c}_{\zeta^2}^2\lambda T)\lambda T\mathfrak{L}(m)^2\int_{-\infty}^{\infty}\gamma(x)^4 f(x)dx.\end{aligned}$$

So,

$$\mathbb{E}\left[\left(\sum_{j=1}^{m}\nu_N(\varphi_j)^2\right)^2\right] \leqslant \mathfrak{c}_2\frac{m\mathfrak{L}(m)^2}{N^2}$$

with

$$\mathfrak{c}_2 = \frac{8\mathfrak{b}_4}{T^4}\left(T^2\int_{-\infty}^{\infty}\sigma(x)^4 f(x)dx + 2(\mathfrak{c}_{\zeta^4}+\mathfrak{c}_{\zeta^2}^2\lambda T)\lambda T\int_{-\infty}^{\infty}\gamma(x)^4 f(x)dx\right).$$

Therefore, there exists a constant $\mathfrak{c}_3 > 0$, not depending on m and N, such that

$$\mathfrak{c}_\Lambda\frac{NT}{\log(NT)\mathfrak{L}(m)}\mathbb{E}\left[\left(\sum_{j=1}^{m}\nu_N(\varphi_j)^2\right)^2\right]^{\frac{1}{2}} \leqslant \mathfrak{c}_3 m^{\frac{1}{2}}.$$

Step 2. First, let us show that $\Phi_{m,\sigma}$ is a positive semidefinite symmetric matrix. Indeed, for any $y \in \mathbb{R}^m$,

$$y^*\Phi_{m,\sigma}y = \frac{1}{NT}\mathbb{E}\left[\left(\sum_{i=1}^{N}\int_0^T \tau_y(X_s^i)(\sigma(X_s^i)dW_s^i + \gamma(X_s^i)d\mathfrak{Z}_s^i)\right)^2\right] \geqslant 0$$

with

$$\tau_y = \sum_{j=1}^{m} y_j \varphi_j.$$

By the isometry property of Itô's integral and the isometry type property of the stochastic integral with respect to $\mathfrak{Z}$,

$$\begin{aligned}
y^*\Phi_{m,\sigma}y &= \frac{1}{T}\mathbb{E}\left[\left(\int_0^T \tau_y(X_s)\sigma(X_s)dW_s\right)^2\right] \\
&\quad + \frac{1}{T}\mathbb{E}\left[\left(\int_0^T \tau_y(X_s)\gamma(X_s)d\mathfrak{Z}_s\right)^2\right] \\
&= \frac{1}{T}\int_0^T \mathbb{E}(\tau_y(X_s)^2\sigma(X_s)^2)ds + \frac{\mathfrak{c}_{\zeta^2}\lambda}{T}\int_0^T \mathbb{E}(\tau_y(X_s)^2\gamma(X_s)^2)ds \\
&\leqslant (\|\sigma\|_\infty^2 + \mathfrak{c}_{\zeta^2}\lambda\|\gamma\|_\infty^2)\underbrace{\int_{-\infty}^{\infty}\left(\sum_{j=1}^{m} y_j\varphi_j(x)\right)^2 f(x)dx}_{=\|\Psi_m^{1/2}y\|_{2,m}^2}. \qquad (4.3)
\end{aligned}$$

Therefore, since $\Phi_{m,\sigma}$ is a positive semidefinite symmetric matrix, by Proposition 3.1 (4, 6) and Inequality (4.3),

$$\begin{aligned}
\frac{1}{m}\text{trace}(\Psi_m^{-1}\Phi_{m,\sigma}) &\leqslant \|\Psi_m^{-\frac{1}{2}}\Phi_{m,\sigma}\Psi_m^{-\frac{1}{2}}\|_{\text{op}} \\
&= \sup_{y:\|\Psi_m^{1/2}y\|_{2,m}=1} y^*\Phi_{m,\sigma}y \leqslant \|\sigma\|_\infty^2 + \mathfrak{c}_{\zeta^2}\lambda\|\gamma\|_\infty^2.
\end{aligned}$$

Step 3 (conclusion). By the two previous steps,

$$\begin{aligned}
\mathbb{E}(\|\widetilde{b}_m - b_0^I\|_N^2) &\leqslant \min_{\tau\in\mathcal{S}_m}\|\tau - b_0^I\|_f^2 \\
&\quad + \frac{2m}{NT}(\|\sigma\|_\infty^2 + \mathfrak{c}_{\zeta^2}\lambda\|\gamma\|_\infty^2) + \mathfrak{c}_3 m^{\frac{1}{2}}\mathbb{P}(\Omega_m^c)^{\frac{1}{2}} + \mathfrak{c}_1\mathbb{P}(\Lambda_m^c)^{\frac{1}{2}}.
\end{aligned}$$

Under Assumption 3.1 with $p \geqslant 2$, by Propositions 3.5 and 3.7, there exists a constant $\mathfrak{c}_4 > 0$, not depending on m and N, such that

$$\mathbb{E}(\|\widetilde{b}_m - b_0^I\|_N^2) \leqslant \min_{\tau\in\mathcal{S}_m}\|\tau - b_0^I\|_f^2 + \mathfrak{c}_4\frac{m}{N}.$$

□

As in Theorem 3.1, the order of the *bias term*

$$\min_{\tau\in\mathcal{S}_m}\|\tau - b_0^I\|_f^2,$$

as well as $\mathfrak{L}(m)$ and $\|\Psi_m^{-1}\|_{\text{op}}$ depend on the φ_j's. For instance, assume that $I = \mathbb{R}$, and that $\varphi_j = h_{j-1}$ for every $j \in \{1, \dots, m\}$, where $(h_n)_{n\in\mathbb{N}}$ is the Hermite basis: for every $x \in \mathbb{R}$ and $n \in \mathbb{N}$,

$$h_n(x) = (2^n n! \sqrt{\pi})^{-\frac{1}{2}} H_n(x) e^{-\frac{x^2}{2}} \quad \text{with} \quad H_n(x) = (-1)^n e^{x^2} \frac{d^n}{dx^n} e^{-x^2}.$$

On the one hand, $\|\varphi_j\|_\infty \leqslant \pi^{-1/4}$ for every $j \in \{1, \dots, m\}$ (see Abramowitz and Stegun [5]), and then $\mathfrak{L}(m) \leqslant m$. Moreover, by Chen et al. [3], Theorem 1.3, for any $t \in (0, T]$ and $x \in \mathbb{R}$,

$$f_t(x) \geqslant \mathfrak{c}_{0.5,T} \left(t^{-\frac{1}{2}} \exp\left(-\mathfrak{m}_{0.5,T} \frac{(x - x_0)^2}{t} \right) + \frac{t}{(t^{1/2} + |x - x_0|)^{1+\alpha}} \right),$$

where $\mathfrak{c}_{0.5,T}$ and $\mathfrak{m}_{0.5,T}$ are positive constants depending on T but not on t and x. So,

$$\begin{aligned} f_t(x) &\geqslant \frac{\mathfrak{c}_{0.5,T}\, t}{(2T + 4x_0^2 + 4x^2)^{(1+\alpha)/2}} \\ &\geqslant \frac{\mathfrak{c}_{0.5,T}\, t}{(4 \vee (2T + 4x_0^2))^{(1+\alpha)/2}} \cdot \frac{1}{(1 + x^2)^{(1+\alpha)/2}}, \end{aligned}$$

and then

$$f(x) \geqslant \frac{\mathfrak{c}_1}{(1 + x^2)^{(1+\alpha)/2}} \quad \text{with} \quad \mathfrak{c}_1 = \frac{\mathfrak{c}_{0.5,T}\, T}{2(4 \vee (2T + 4x_0^2))^{(1+\alpha)/2}}.$$

By assuming that $\alpha \in [1, 2)$, and by Comte and Genon-Catalot [6], Proposition 9, there exists a constant $\mathfrak{c}_2 > 0$, not depending on m, such that

$$\|\Psi_m^{-1}\|_{\text{op}} \leqslant \mathfrak{c}_2 m^{\frac{1+\alpha}{2}}.$$

Therefore, Assumption 3.1 is fulfilled by any

$$m \leqslant \left(\frac{\overline{\mathfrak{c}}_\Lambda}{\mathfrak{c}_2 \vee 1} \cdot \frac{NT}{\log(NT)} \right)^{\frac{2}{3+\alpha}}.$$

On the other hand, consider the Hermite-Sobolev ball

$$\mathbb{W}_{\mathrm{H}}^\beta(\delta) = \left\{ \varphi \in \mathbb{L}^2(\mathbb{R}, dx) : \sum_{n=0}^{\infty} n^\beta \langle \varphi, h_n \rangle^2 \leqslant \delta \right\}$$

with $\beta > (3 + \alpha)/2 - 1$ and $\delta > 0$, and assume that $b_0 \in \mathbb{W}_{\mathrm{H}}^\beta(\delta)$. By Belomestny et al. [7],

$$\|\Pi_m(b_0) - b_0\|^2 \leqslant \delta m^{-\beta}.$$

Since f is upper bounded on $\mathbb{R}$ by a constant $\overline{\mathfrak{m}} > 0$ thanks to Inequality (4.2),

$$\min_{\tau \in S_m} \|\tau - b_0\|_f^2 \leqslant \overline{\mathfrak{m}} \|\Pi_m(b_0) - b_0\|^2$$
$$\leqslant \mathfrak{c}_3 m^{-\beta} \quad \text{with} \quad \mathfrak{c}_3 = \overline{\mathfrak{m}}\delta.$$

In conclusion, by Theorem 4.1, there exists a constant $\overline{\mathfrak{c}}_{4.1} > 0$, not depending on m and N, such that

$$\mathbb{E}(\|\widetilde{b}_m - b_0\|_N^2) \leqslant \overline{\mathfrak{c}}_{4.1}\left(m^{-\beta} + \frac{m}{N}\right),$$

and then the bias-variance tradeoff is reached by (the risk bound on) $\widetilde{b}_m$ for m of order $N^{1/(1+\beta)}$.

4.1.3 Model Selection

As in Chap. 3, the dimension m of S_m needs to be selected from data because the one for which $\widetilde{b}_m$ reaches the bias-variance tradeoff depends on some unknown regularity parameters of b_0. In order to introduce an appropriate model selection criterion, throughout this section, the φ_j's fulfill Assumption 3.4 as in Sect. 3.6. Moreover, in order to control the *big jumps* of X, the Lévy measure π_λ needs to fulfill the following assumption.

Assumption 4.1 The Lévy measure π_λ is sub-exponential: there exist $\mathfrak{a}, \mathfrak{b} > 0$ such that, for every $x > 1$,

$$\pi_\lambda((-x, x)^c) \leqslant \mathfrak{a} e^{-\mathfrak{b}|x|}.$$

Now, let us consider

$$\widehat{m} = \arg\min_{m \in \widehat{\mathcal{M}}_N} \{-\|\widehat{b}_m\|_N^2 + \text{pen}(m)\} \tag{4.4}$$

where

$$\text{pen}(m) = \mathfrak{c}_{\text{cal}} \frac{m}{N};\ \forall m \in \{1, \dots, N_T\},$$

the constant $\mathfrak{c}_{\text{cal}} > 0$ needs to be calibrated in practice,

$$\widehat{\mathcal{M}}_N = \left\{m \in \{1, \dots, N_T\} : \mathfrak{c}_\varphi^2 m(\|\widehat{\Psi}_m^{-1}\|_{\text{op}}^2 \vee 1) \leqslant \mathfrak{d}\frac{NT}{\log(NT)}\right\}$$

and

$$\mathfrak{d} = \min\left\{\frac{1}{8\mathfrak{c}_\varphi^2 T(\|f\|_\infty + \mathfrak{c}_{\Lambda,\varphi})(1+p)}, \frac{\overline{\mathfrak{c}}_\Lambda}{4}\right\} \quad \text{with} \quad \mathfrak{c}_{\Lambda,\varphi} = \frac{\sqrt{\overline{\mathfrak{c}}_\Lambda/4}}{3\mathfrak{c}_\varphi}.$$

Consider also the theoretical counterpart

$$\mathcal{M}_N = \left\{ m \in \{1, \dots, N_T\} : \mathfrak{c}_{\varphi}^2 m(\|\Psi_m^{-1}\|_{\text{op}}^2 \vee 1) \leqslant \overline{\mathfrak{d}} \frac{NT}{\log(NT)} \right\} \quad \text{of} \quad \widehat{\mathcal{M}}_N,$$

where $\overline{\mathfrak{d}} = \mathfrak{d}/4$.

As in Sect. 3.6, in order to establish a risk bound on the adaptive estimator $\widehat{b}_{\widehat{m}}$, the deterministic set $\mathcal{M}_N$ needs to be almost surely contained in its theoretical counterpart when $N \to \infty$. To that purpose, let us introduce the event

$$\Xi_{N,\Gamma} = \{\mathcal{M}_{N,\gamma} \subset \widehat{\mathcal{M}}_{N,\Gamma} \subset \mathfrak{M}_{N,\gamma}\},$$

where

- $\widehat{\mathcal{M}}_{N,\Gamma} = \left\{ m \in \{1, \dots, N_T\} : \mathfrak{c}_{\varphi}^2 m(\|\widehat{\Psi}_{m,\Gamma}^{-1}\|_{\text{op}}^2 \vee 1) \leqslant \mathfrak{d} \frac{NT}{\log(NT)} \right\}$,
- $\mathcal{M}_{N,\gamma} = \left\{ m \in \{1, \dots, N_T\} : \mathfrak{c}_{\varphi}^2 m(\|\Psi_{m,\gamma}^{-1}\|_{\text{op}}^2 \vee 1) \leqslant \overline{\mathfrak{d}} \frac{NT}{\log(NT)} \right\}$,
- and $\mathfrak{M}_{N,\gamma} = \left\{ m \in \{1, \dots, N_T\} : \mathfrak{c}_{\varphi}^2 m(\|\Psi_{m,\gamma}^{-1}\|_{\text{op}}^2 \vee 1) \leqslant \widetilde{\mathfrak{d}} \frac{NT}{\log(NT)} \right\}$ with $\widetilde{\mathfrak{d}} = 4\mathfrak{d}$.

The following proposition provides a suitable bound on $\mathbb{P}(\Xi_{N,\Gamma}^c)$ when $\Gamma^1, \dots, \Gamma^N$ are independent.

Proposition 4.2 *Under Assumption 3.4, there exists a constant $\mathfrak{c}_{4.2} > 0$, not depending on N, such that*

$$\mathbb{P}(\Xi_{N,\Gamma}^c) \leqslant \frac{\mathfrak{c}_{4.2}}{N^{p-1}}.$$

Proof First of all, note that

$$\Xi_{N,\Gamma}^c = \{\mathcal{M}_{N,\gamma} \not\subset \widehat{\mathcal{M}}_{N,\Gamma}\} \cup \{\widehat{\mathcal{M}}_{N,\Gamma} \not\subset \mathfrak{M}_{N,\gamma}\}.$$

The proof of Proposition 4.2 is dissected in three steps. Step 1 provides a bound on $\mathbb{P}(\mathcal{M}_{N,\gamma} \not\subset \widehat{\mathcal{M}}_{N,\Gamma})$, Step 2 a bound on $\mathbb{P}(\|\widehat{\Psi}_{m,\Gamma} - \Psi_{m,\gamma}\|_{\text{op}} > \delta)$ for every $m \in \{1, \dots, N_T\}$ and $\delta > 0$, and then Step 3 a bound on $\mathbb{P}(\widehat{\mathcal{M}}_{N,\Gamma} \not\subset \mathfrak{M}_{N,\gamma})$.

Step 1. On $\{\mathcal{M}_{N,\gamma} \not\subset \widehat{\mathcal{M}}_{N,\Gamma}\}$, there exists $m \in \{1, \dots, N_T\}$ such that

$$\mathfrak{c}_{\varphi}^2 m(\|\Psi_{m,\gamma}^{-1}\|_{\text{op}}^2 \vee 1) \leqslant \overline{\mathfrak{d}} \frac{NT}{\log(NT)} \quad \text{and} \quad \mathfrak{c}_{\varphi}^2 m(\|\widehat{\Psi}_{m,\Gamma}^{-1}\|_{\text{op}}^2 \vee 1) > \mathfrak{d} \frac{NT}{\log(NT)}.$$

The first inequality is equivalent to

$$\mathfrak{c}_{\varphi}^2 m\|\Psi_{m,\gamma}^{-1}\|_{\text{op}}^2 \leqslant \overline{\mathfrak{d}} \frac{NT}{\log(NT)} \quad \text{and} \quad \mathfrak{c}_{\varphi}^2 m \leqslant \overline{\mathfrak{d}} \frac{NT}{\log(NT)},$$

and then the second one leads to

$$\mathfrak{d}\frac{NT}{\log(NT)} < \mathfrak{c}_{\varphi}^2 m\|\widehat{\Psi}_{m,\Gamma}^{-1}\|_{\text{op}}^2 \leqslant 2\mathfrak{c}_{\varphi}^2 m(\|\widehat{\Psi}_{m,\Gamma}^{-1} - \Psi_{m,\gamma}^{-1}\|_{\text{op}}^2 + \|\Psi_{m,\gamma}^{-1}\|_{\text{op}}^2)$$
$$\leqslant 2\mathfrak{c}_{\varphi}^2 m\|\widehat{\Psi}_{m,\Gamma}^{-1} - \Psi_{m,\gamma}^{-1}\|_{\text{op}}^2 + 2\overline{\mathfrak{d}}\frac{NT}{\log(NT)}.$$

So, since $\mathfrak{d} - 2\overline{\mathfrak{d}} = 2\overline{\mathfrak{d}}$, and by the definition of $\mathcal{M}_{N,\gamma}$,

$$\{\mathcal{M}_{N,\gamma} \not\subset \widehat{\mathcal{M}}_{N,\Gamma}\} \subset \bigcup_{m\in\mathcal{M}_{N,\gamma}} \left\{\overline{\mathfrak{d}}\frac{NT}{\log(NT)} \leqslant \mathfrak{c}_{\varphi}^2 m\|\widehat{\Psi}_{m,\Gamma}^{-1} - \Psi_{m,\gamma}^{-1}\|_{\text{op}}^2\right\}$$
$$\subset \bigcup_{m\in\mathcal{M}_{N,\gamma}} \{\|\Psi_{m,\gamma}^{-1}\|_{\text{op}} < \|\widehat{\Psi}_{m,\Gamma}^{-1} - \Psi_{m,\gamma}^{-1}\|_{\text{op}}\} \subset \bigcup_{m\in\mathcal{M}_{N,\gamma}} \Omega_{m,\Gamma}^c.$$

Therefore, since $\overline{\mathfrak{d}} \leqslant \overline{\mathfrak{c}}_{\Lambda}$, by Proposition 3.6,

$$\mathbb{P}(\mathcal{M}_{N,\gamma} \not\subset \widehat{\mathcal{M}}_{N,\Gamma}) \leqslant \sum_{m\in\mathcal{M}_{N,\gamma}} \mathbb{P}(\Omega_{m,\Gamma}^c) \leqslant \frac{\mathfrak{c}_1}{N^{p-1}},$$

where $\mathfrak{c}_1 > 0$ is a constant not depending on N.

Step 2. For any $m \in \{1, \dots, N_T\}$, let us recall that

$$\widehat{\Psi}_{m,\Gamma} = \frac{1}{N}\sum_{i=1}^{N} \Psi_m(\Gamma^i),$$

and that $\|\Psi_m(h)\|_{\text{op}} \leqslant \mathfrak{L}(m)$ for every piecewise continuous function $h : [0, T] \to \mathbb{R}$ (see Inequality (3.4)). Then, for $A = N^{-1}(\Psi_m(\Gamma) - \mathbb{E}(\Psi_m(\Gamma)))$,

$$\mathbb{E}(A) = 0 \quad \text{and} \quad \|A\|_{\text{op}} \leqslant R = 2\frac{\mathfrak{L}(m)}{N}.$$

Moreover,

$$\|\mathbb{E}(A^2)\|_{\text{op}} = \sup_{y:\|y\|_{2,m}=1} \mathbb{E}(\|Ay\|_{2,m}^2) = \frac{1}{N^2}\sup_{y:\|y\|_{2,m}=1}\sum_{j=1}^{m} \text{var}([\Psi_m(\Gamma)y]_j)$$

and, for every $j \in \{1, \dots, m\}$ and $y \in \partial B_{2,m}(0, 1)$,

$$\text{var}([\Psi_m(\Gamma)y]_j) \leqslant \mathbb{E}\left[\left(\sum_{\ell=1}^{m} y_\ell \cdot \frac{1}{T}\int_0^T \varphi_j(\Gamma_s)\varphi_\ell(\Gamma_s)ds\right)^2\right]$$

$$\leqslant \int_{-\infty}^{\infty}\left(\sum_{\ell=1}^{m}\varphi_j(x)\varphi_\ell(x)y_\ell\right)^2 f(x)dx \leqslant \|f\|_\infty\|\varphi_j\|_\infty^2 \underbrace{\left\|\sum_{\ell=1}^{m} y_\ell\varphi_\ell\right\|^2}_{=1}.$$

So,

$$\sigma^2 = N\|\mathbb{E}(A^2)\|_{\rm op} \leqslant \|f\|_\infty \frac{\mathfrak{L}(m)}{N}.$$

Therefore, by the matrix Bernstein inequality (see Proposition 3.3), for every $\delta \geqslant 0$,

$$\begin{aligned}\mathbb{P}(\|\widehat{\Psi}_{m,\Gamma} - \Psi_{m,\gamma}\|_{\rm op} \geqslant \delta) &\leqslant m\exp\left(-\frac{\delta^2/2}{\sigma^2 + R\delta/3}\right)\\ &\leqslant m\exp\left(-\frac{N\delta^2/2}{\mathfrak{L}(m)(\|f\|_\infty + 2\delta/3)}\right).\end{aligned}$$

Step 3. On $\{\widehat{\mathcal{M}}_{N,\Gamma} \not\subset \mathfrak{M}_{N,\gamma}\}$, there exists $m \in \{1,\dots,N_T\}$ such that

$$\mathfrak{c}_\varphi^2 m(\|\widehat{\Psi}_{m,\Gamma}^{-1}\|_{\rm op}^2 \vee 1) \leqslant \mathfrak{d}\frac{NT}{\log(NT)} \quad\text{and}\quad \mathfrak{c}_\varphi^2 m(\|\Psi_{m,\gamma}^{-1}\|_{\rm op}^2 \vee 1) > \widetilde{\mathfrak{d}}\frac{NT}{\log(NT)}.$$

The first inequality is equivalent to

$$\mathfrak{c}_\varphi^2 m\|\widehat{\Psi}_{m,\Gamma}^{-1}\|_{\rm op}^2 \leqslant \mathfrak{d}\frac{NT}{\log(NT)} \quad\text{and}\quad \mathfrak{c}_\varphi^2 m \leqslant \mathfrak{d}\frac{NT}{\log(NT)},$$

and then the second one leads to

$$\begin{aligned}\widetilde{\mathfrak{d}}\frac{NT}{\log(NT)} &< \mathfrak{c}_\varphi^2 m\|\Psi_{m,\gamma}^{-1}\|_{\rm op}^2 \leqslant 2\mathfrak{c}_\varphi^2 m(\|\Psi_{m,\gamma}^{-1} - \widehat{\Psi}_{m,\Gamma}^{-1}\|_{\rm op}^2 + \|\widehat{\Psi}_{m,\Gamma}^{-1}\|_{\rm op}^2)\\ &\leqslant 2\mathfrak{c}_\varphi^2 m\|\Psi_{m,\gamma}^{-1} - \widehat{\Psi}_{m,\Gamma}^{-1}\|_{\rm op}^2 + 2\mathfrak{d}\frac{NT}{\log(NT)}.\end{aligned}$$

Moreover, for every $m \in \{1,\dots,N_T\}$, by interchanging $\widehat{\Psi}_{m,\Gamma}$ and $\Psi_{m,\gamma}$ in the proof of Lemma 3.3,

$$\begin{aligned}\{\|\Psi_{m,\gamma}^{-1} - \widehat{\Psi}_{m,\Gamma}^{-1}\|_{\rm op} > \|\widehat{\Psi}_{m,\Gamma}^{-1}\|_{\rm op}\} &\subset \left\{\|\widehat{\Psi}_{m,\Gamma}^{-\frac{1}{2}}\Psi_{m,\gamma}\widehat{\Psi}_{m,\Gamma}^{-\frac{1}{2}} - I\|_{\rm op} > \frac{1}{2}\right\}\\ &\subset \left\{\|\widehat{\Psi}_{m,\Gamma} - \Psi_{m,\gamma}\|_{\rm op} > \frac{1}{2}\|\widehat{\Psi}_{m,\Gamma}^{-1}\|_{\rm op}^{-1}\right\}.\end{aligned}$$

So, since $\widetilde{\mathfrak{d}} - 2\mathfrak{d} = 2\mathfrak{d}$,

$\{\widehat{\mathcal{M}}_{N,\Gamma} \not\subset \mathfrak{M}_{N,\gamma}\}$

$$\subset \bigcup_{\mathfrak{c}_\varphi^2 m \leqslant \mathfrak{d}\frac{NT}{\log(NT)}} \left\{\mathfrak{c}_\varphi^2 m\|\widehat{\Psi}_{m,\Gamma}^{-1}\|_{\mathrm{op}}^2 \leqslant \mathfrak{d}\frac{NT}{\log(NT)} < \mathfrak{c}_\varphi^2 m\|\Psi_{m,\gamma}^{-1} - \widehat{\Psi}_{m,\Gamma}^{-1}\|_{\mathrm{op}}^2\right\}$$

$$\subset \bigcup_{\mathfrak{c}_\varphi^2 m \leqslant \mathfrak{d}\frac{NT}{\log(NT)}} \left\{\|\widehat{\Psi}_{m,\Gamma} - \Psi_{m,\gamma}\|_{\mathrm{op}} > \frac{1}{2}\left(\frac{m\log(NT)}{\mathfrak{d}NT}\right)^{\frac{1}{2}}\right\}.$$

Therefore, by Step 2 and Assumption 3.4,

$$\begin{aligned}
\mathbb{P}(\widehat{\mathcal{M}}_{N,\Gamma} \not\subset \mathfrak{M}_{N,\gamma}) &\leqslant \sum_{\mathfrak{c}_\varphi^2 m \leqslant \mathfrak{d}\frac{NT}{\log(NT)}} m\exp\left(-\frac{m/(8\mathfrak{d}T)\log(NT)}{\mathfrak{L}(m)(\|f\|_\infty + \mathfrak{d}^{1/2}/(3\mathfrak{c}_\varphi))}\right)\\
&\leqslant \sum_{\mathfrak{c}_\varphi^2 m \leqslant \mathfrak{d}\frac{NT}{\log(NT)}} m\exp\left(-\frac{\log(NT)}{8\mathfrak{c}_\varphi^2\mathfrak{d}T(\|f\|_\infty + \mathfrak{c}_{\Lambda,\varphi})}\right)\\
&\leqslant \frac{1}{(NT)^{p+1}} \sum_{\mathfrak{c}_\varphi^2 m \leqslant \mathfrak{d}\frac{NT}{\log(NT)}} m \leqslant \frac{\mathfrak{c}_2}{N^{p-1}},
\end{aligned}$$

where $\mathfrak{c}_2 > 0$ is a constant not depending on N. □

The following theorem, which is the main result of this section, provides a risk bound on the adaptive estimator $\widehat{b}_{\widehat{m}}$.

Theorem 4.2 *Under Assumptions 3.4 and 4.1, there exists a constant* $\mathfrak{c}_{3.2} > 0$*, not depending on* N*, such that*

$$\mathbb{E}(\|\widehat{b}_{\widehat{m}} - b_0^I\|_N^2) \leqslant \mathfrak{c}_{3.2}\left(\min_{m\in\mathcal{M}_N}\left\{\min_{\tau\in\mathcal{S}_m}\|\tau - b_0^I\|_f^2 + \frac{m}{N}\right\} + \frac{1}{N}\right).$$

Proof Let Ω_N be the event defined by

$$\Omega_N = \bigcap_{m\in\mathfrak{M}_N} \Omega_m$$

and, for every $m, m' \in \mathcal{M}_N$, consider

$$\rho_N(m) = \mathbb{E}\left(\left(\left[\sup_{\tau\in\mathcal{B}_{m,\widehat{m}}} |\nu_N(\tau)|\right]^2 - p(m,\widehat{m})\right)_+ \mathbf{1}_{\Xi_N\cap\Omega_N}\right)$$

and

$$p(m, m') = \frac{\mathfrak{c}_{\mathrm{cal}}}{8}\cdot\frac{m\vee m'}{N}.$$

Since the definition of the empirical projection $\widehat{\Pi}_m$ remains the same as in Chap. 3, and since

exists a unique α-Hölder continuous function $J_{h,x} : [0, T] \to \mathbb{R}$ such that, for any $s, t \in [0, T]$ satisfying $s < t$, and every sequence $(D_n)_{n\in\mathbb{N}}$ of dissections of $[s, t]$ satisfying $\pi(D_n) \to 0$ when $n \to \infty$,

$$\lim_{n\to\infty} |J_{h,x}(t) - J_{h,x}(s) - J_{h,x,D_n}(s, t)| = 0.$$

Young's integral on $[s, t]$ of h with respect to x is defined by

$$\int_s^t h(u)dx(u) = J_{h,x}(t) - J_{h,x}(s).$$

See Friz and Victoir [10], Theorem 6.8 for a proof. Now, let us state the change of variable formula for Young's integral.

Proposition 4.3 *Consider $F \in C^1(\mathbb{R})$ such that F' is Lipschitz continuous, and let x be an α-Hölder continuous function from $[0, T]$ into $\mathbb{R}$ with $\alpha \in (1/2, 1]$. Then, for every $t \in (0, T]$,*

$$F(x(t)) = F(x(0)) + \int_0^t F'(x(s))dx(s).$$

Note that Proposition 4.3 is a consequence of the Taylor formula with integral remainder.

For any $\alpha \in (1/2, \mathrm{H})$, the paths of B are α-Hölder continuous (see Nualart [11], Sect. 5.1), and so are those of X by Eq. (4.8). Then, for every process $H = (H_t)_{t\in[0,T]}$ having β-Hölder continuous paths from $[0, T]$ into $\mathbb{R}$ with $\beta \in (0, 1]$ such that $\alpha + \beta > 1$, by Theorem 4.3, one may define the pathwise integral on $[0, T]$ of H with respect to B (resp. X) by

$$\left(\int_0^T H_s dB_s\right)(\omega) = \int_0^T H_s(\omega)dB_s(\omega)$$
$$\left(\text{resp. } \left(\int_0^T H_s dX_s\right)(\omega) = \int_0^T H_s(\omega)dX_s(\omega)\right); \forall \omega \in \Omega.$$

To conclude this section, note that the pathwise integral has a role to play in the sequel, but not for statistical purposes directly, especially because the pathwise integral with respect to B is not centered and there is no sharp enough control of its fourth-order moment.

4.2.1.2 The Skorokhod Integral

Let $\langle \cdot, \cdot \rangle_{\mathcal{H}}$ be the inner product defined by

$$\langle h, \eta \rangle_{\mathcal{H}} := \alpha_{\mathrm{H}} \int_0^T \int_0^T h(s)\eta(t)|t-s|^{2\mathrm{H}-2}dsdt \quad \text{with} \quad \alpha_{\mathrm{H}} = \mathrm{H}(2\mathrm{H}-1),$$

and consider the reproducing kernel Hilbert space $\mathcal{H} = \{h : \|h\|_{\mathcal{H}} < \infty\}$ of B, where $\|.\|_{\mathcal{H}}$ is the norm associated to $\langle \cdot, \cdot \rangle_{\mathcal{H}}$. Consider also the isonormal Gaussian process $(\mathbf{B}(h))_{h \in \mathcal{H}}$ defined by

$$\mathbf{B}(h) = \int_0^T h(s)dB_s; h \in \mathcal{H},$$

which is the Wiener integral of h with respect to B on $[0, T]$.

Notation. The space of all the smooth functions $\varphi : \mathbb{R}^n \to \mathbb{R}$ ($n \in \mathbb{N}^*$) such that φ and all its partial derivatives have polynomial growth is denoted by $C_{\mathrm{p}}^{\infty}(\mathbb{R}^n; \mathbb{R})$.

Definition 4.2 The Malliavin derivative of the smooth functional

$$F = \varphi(\mathbf{B}(h_1), \dots, \mathbf{B}(h_n))$$

with $n \in \mathbb{N}^*$, $\varphi \in C_{\mathrm{p}}^{\infty}(\mathbb{R}^n; \mathbb{R})$ and $h_1, \dots, h_n \in \mathcal{H}$, is the $\mathcal{H}$-valued random variable

$$\mathbf{D}F = \sum_{k=1}^{n} \partial_k \varphi(\mathbf{B}(h_1), \dots, \mathbf{B}(h_n))h_k.$$

Proposition 4.4 *The map* $\mathbf{D}$ *is closable from* $\mathbb{L}^2(\Omega; \mathbb{R})$ *into* $\mathbb{L}^2(\Omega; \mathcal{H})$. *Its domain in* $\mathbb{L}^2(\Omega; \mathbb{R})$, *denoted by* $\mathbb{D}^{1,2}$, *is the closure of the smooth functionals space for the norm* $\|.\|_{1,2}$ *defined by*

$$\|F\|_{1,2}^2 = \mathbb{E}(F^2) + \mathbb{E}(\|\mathbf{D}F\|_{\mathcal{H}}^2).$$

The Malliavin derivative of $F \in \mathbb{D}^{1,2}$ *at time* $s \in [0, T]$ *is denoted by* $\mathbf{D}_s F$.

See Nualart [11], Proposition 1.2.1 for a proof. Let us state the chain rule for the Malliavin derivative.

Proposition 4.5 *Consider* $F \in \mathbb{D}^{1,2}$, *and* $\varphi \in C^1(\mathbb{R})$ *such that* φ' *is bounded. Then,*

$$\mathbf{D}(\varphi(F)) = \varphi'(F)\mathbf{D}F.$$

See Nualart [11], Proposition 1.2.3 for a proof.

Definition 4.3 The adjoint δ of the Malliavin derivative $\mathbf{D}$ is the divergence operator. The domain of δ is denoted by dom(δ), and $H \in \mathrm{dom}(\delta)$ if and only if there exists a constant $\mathfrak{c}_H > 0$ such that, for every $F \in \mathbb{D}^{1,2}$,

$$|\mathbb{E}(\langle \mathbf{D}F, H \rangle_{\mathcal{H}})| \leqslant \mathfrak{c}_H \mathbb{E}(F^2)^{\frac{1}{2}}.$$

Let $\mathcal{S}$ be the space of the smooth functionals presented in Definition 4.2, and consider $\mathbb{D}^{1,2}(\mathcal{H})$; the closure of

$$\mathcal{S}_{\mathcal{H}} = \left\{ \sum_{j=1}^{n} F_j h_j ; h_1, \dots, h_n \in \mathcal{H}, F_1, \dots, F_n \in \mathcal{S} \right\}$$

for the norm $\|.\|_{1,2,\mathcal{H}}$ defined by

$$\|H\|_{1,2,\mathcal{H}}^2 = \mathbb{E}(\|H\|_{\mathcal{H}}^2) + \mathbb{E}(\|\mathbf{D}H\|_{\mathcal{H}\otimes\mathcal{H}}^2).$$

By Nualart [11], Proposition 1.3.1,

$$\mathbb{D}^{1,2}(\mathcal{H}) \subset \text{dom}(\delta).$$

Let us state the isometry type property for the divergence operator on $\mathbb{D}^{1,2}(\mathcal{H})$.

Proposition 4.6 *For every $H, \overline{H} \in \mathbb{D}^{1,2}(\mathcal{H})$,*

$$\mathbb{E}(\delta(H)\delta(\overline{H})) = \alpha_{\mathrm{H}} \int_0^T \int_0^T \mathbb{E}(H_s \overline{H}_t)|t-s|^{2\mathrm{H}-2} ds dt + R_{H,\overline{H}}$$

with

$$R_{H,\overline{H}} = \alpha_{\mathrm{H}}^2 \int_{[0,T]^4} \mathbb{E}(\mathbf{D}_{\overline{u}} H_v \mathbf{D}_{\overline{v}} \overline{H}_u)|u-\overline{u}|^{2\mathrm{H}-2}|v-\overline{v}|^{2\mathrm{H}-2} d\overline{u} d\overline{v} du dv.$$

The reader may refer to Biagini et al. [12], Theorem 3.11.1 for a proof.

For any process $H = (H_s)_{s\in[0,T]}$ and any $t \in (0, T]$, if $H\mathbf{1}_{[0,t]} \in \text{dom}(\delta)$, then its Skorokhod integral with respect to B is defined on $[0, t]$ by

$$\int_0^t H_s \delta B_s = \delta(H\mathbf{1}_{[0,t]}),$$

and its Skorokhod integral with respect to the solution X of Eq. (4.8) is defined by

$$\int_0^t H_s \delta X_s = \int_0^t H_s b_0(X_s) ds + \sigma \int_0^t H_s \delta B_s.$$

Note that since δ is the adjoint of the Malliavin derivative $\mathbf{D}$, the Skorokhod integral on $[0, t]$ of H with respect to B is a centered random variable. Indeed,

$$\mathbb{E}\left(\int_0^t H_s \delta B_s\right) = \mathbb{E}(1 \cdot \delta(H\mathbf{1}_{[0,t]})) = \mathbb{E}(\langle \mathbf{D}(1), H\mathbf{1}_{[0,t]}\rangle_{\mathcal{H}}) = 0 \qquad (4.9)$$

Now, let us state a suitable relationship between Skorokhod's integral and the pathwise integral with respect to X.

Proposition 4.7 *For every $\varphi \in C^1(\mathbb{R})$ of bounded derivative, $(\varphi(X_t))_{t\in[0,T]}$ belongs to $\mathbb{D}^{1,2}(\mathcal{H})$ and*

$$\int_0^T \varphi(X_s)\delta X_s = \int_0^T \varphi(X_s)dX_s - \alpha_{\mathrm{H}}\sigma^2 \int_0^T \int_0^t \varphi'(X_t)\exp\left(\int_s^t b_0'(X_u)du\right)|t-s|^{2\mathrm{H}-2}dsdt$$

Proof For any $s, t \in [0, T]$ such that $s < t$, since

$$\mathbf{D}_s B_t = \mathbf{D}_s(\mathbf{B}(\mathbf{1}_{[0,t]})) = \mathbf{1}_{[0,t]}(s),$$

by Eq. (4.8) and Proposition 4.5,

$$\mathbf{D}_s X_t = \int_0^t b_0'(X_u)\mathbf{D}_s X_u du + \sigma\mathbf{1}_{[0,t]}(s).$$

Therefore,

$$\mathbf{D}_s X_t = \sigma\mathbf{1}_{[0,t]}(s)\exp\left(\int_s^t b_0'(X_u)du\right)$$

and, by Proposition 4.5 and Nualart [11], Proposition 5.2.3,

$$\begin{aligned}\int_0^T \varphi(X_s)\delta X_s &= \int_0^T \varphi(X_s)dX_s - \alpha_{\mathrm{H}}\sigma \int_0^T \int_0^T \mathbf{D}_s(\varphi(X_t))|t-s|^{2\mathrm{H}-2}dsdt \\ &= \int_0^T \varphi(X_s)dX_s \\ &\quad - \alpha_{\mathrm{H}}\sigma^2 \int_0^T \int_0^t \varphi'(X_t)\exp\left(\int_s^t b_0'(X_u)du\right)|t-s|^{2\mathrm{H}-2}dsdt.\end{aligned}$$

□

Finally, the following proposition provides (in particular) a suitable control of the fourth-order moment of Skorokhod's integral with respect to B.

Proposition 4.8 *Consider $p > 1/\mathrm{H}$ and*

$$M = \sup_{x\in\mathbb{R}} b_0'(x).$$

There exists a constant $\mathfrak{c}_{4.8} > 0$, only depending on p, H and σ, such that for every $\varphi \in C^1(\mathbb{R})$ of bounded derivative,

4.2.2.2 On a Computable Approximation of $\widehat{\theta}_N$

The Skorokhod integral, and then $\widehat{\theta}_N$, are not directly computable. However, this section deals with the approximation $\overline{\theta}_N$ of $\widehat{\theta}_N$, which is computable by solving Eq. (4.10). First of all, let us explain why $\overline{\theta}_N$ is defined this way. For every $i \in \{1, \dots, N\}$, by Proposition 4.7,

$$\int_0^T b(X_s^i)\delta X_s^i = \int_0^T b(X_s^i)dX_s^i \\ - \alpha_{\mathrm{H}}\sigma^2 \int_0^T \int_0^t b'(X_t^i)\exp\left(\theta_0 \int_s^t b'(X_u^i)du\right)|t-s|^{2\mathrm{H}-2}dsdt.$$

Then, since $\widehat{\theta}_N$ is a converging estimator of θ_0 as established in Sect. 4.2.2.1,

$$\widehat{\theta}_N - I_N = \Theta_N(\theta_0 - I_N) \approx \Theta_N(\widehat{\theta}_N - I_N),$$

where

$$\Theta_N(\cdot) = -\frac{\alpha_{\mathrm{H}}\sigma^2}{NTD_N}\sum_{i=1}^N \int_0^T \int_0^t b'(X_t^i)\exp\left((\cdot + I_N)\int_s^t b'(X_u^i)du\right)|t-s|^{2\mathrm{H}-2}dsdt$$

and, by the change of variable formula for Young's integral,

$$I_N = \frac{1}{NTD_N}\sum_{i=1}^N \int_0^T b(X_s^i)dX_s^i \\ = \frac{1}{NTD_N}\sum_{i=1}^N (\mathrm{b}(X_T^i) - \mathrm{b}(x_0)) \quad \text{with} \quad \mathrm{b}' = b.$$

This legitimates to consider the estimator $\overline{\theta}_N = I_N + R_N$ of θ_0, where R_N is a fixed point of the map Θ_N. Let us establish that R_N exists and is unique under the condition (4.14) stated below.

Proposition 4.10 *Assume that $b'(\cdot) \leqslant 0$. If*

$$T^{2\mathrm{H}}\frac{M_N}{D_N} \leqslant \frac{\mathfrak{c}}{\overline{\alpha}_{\mathrm{H}}\sigma^2\|b'\|_\infty^2} \tag{4.14}$$

where $\mathfrak{c}$ is a deterministic constant arbitrarily chosen in $(0, 1)$,

$$M_N = e^{\|b'\|_\infty |I_N| T} \quad \text{and} \quad \overline{\alpha}_{\mathrm{H}} = \frac{\alpha_{\mathrm{H}}}{2\mathrm{H}(2\mathrm{H}+1)},$$

then Θ_N is a contraction from $\mathbb{R}_+$ into itself. Therefore, R_N exists and is unique.

Proof Since $b'(\cdot) \leqslant 0$, Θ_N is nonnegative, and in particular $\Theta_N(\mathbb{R}_+) \subset \mathbb{R}_+$. Moreover, by (4.14), for every $r, \overline{r} \in \mathbb{R}_+$,

$$\begin{aligned}
|\Theta_N(\overline{r}) - \Theta_N(r)| &\leqslant \frac{\alpha_{\mathrm{H}}\sigma^2}{NTD_N}\sum_{i=1}^{N}\int_0^T\int_0^t |t-s|^{2\mathrm{H}-2}|b'(X_t^i)| \exp\left(I_N\int_s^t b'(X_u^i)du\right)\\
&\quad \times \left|\exp\left(\overline{r}\int_s^t b'(X_u^i)du\right) - \exp\left(r\int_s^t b'(X_u^i)du\right)\right| dsdt\\
&\leqslant \frac{\alpha_{\mathrm{H}}\sigma^2}{NTD_N}\|b'\|_{\infty}M_N\\
&\quad \times \sum_{i=1}^{N}\int_0^T\int_0^t |t-s|^{2\mathrm{H}-2}\sup_{x\in\mathbb{R}_-} e^x \left|(\overline{r}-r)\int_s^t b'(X_u^i)du\right| dsdt\\
&\leqslant \overline{\alpha}_{\mathrm{H}}\sigma^2\|b'\|_{\infty}^2 T^{2\mathrm{H}}\frac{M_N}{D_N}|\overline{r}-r| \leqslant \mathfrak{c}|\overline{r}-r|.
\end{aligned}$$

So, Θ_N is a contraction from $\mathbb{R}_+$ into $\mathbb{R}_+$, and then R_N exists and is unique by Picard's fixed point theorem.

□

Let us consider $\overline{\theta}_N^{\mathfrak{c}} = \overline{\theta}_N \mathbf{1}_{\Delta_N}$, where

$$\Delta_N = \left\{T^{2\mathrm{H}}\frac{M_N}{D_N} \leqslant \frac{\mathfrak{c}}{\overline{\alpha}_{\mathrm{H}}\sigma^2\|b'\|_{\infty}^2}\right\},$$

and the truncated estimator

$$\overline{\theta}_N^{\mathfrak{c},\mathfrak{d}} = \overline{\theta}_N^{\mathfrak{c}}\mathbf{1}_{D_N \geqslant \mathfrak{d}} \quad \text{with} \quad \mathfrak{d} \in \Delta_f = \left(0, \frac{\|b\|_f^2}{2}\right].$$

The following proposition provides a risk bound on $\overline{\theta}_N^{\mathfrak{c},\mathfrak{d}}$.

Proposition 4.11 *Assume that $b'(\cdot) \leqslant 0$, $\theta_0 > 0$ and that*

$$\frac{2T^{2\mathrm{H}}}{\|b\|_f^2}\exp\left(\frac{2\|b'\|_{\infty}}{\|b\|_f^2}|\mathbb{E}(\mathrm{b}(X_T)) - \mathrm{b}(x_0)|\right) < \frac{\mathfrak{c}}{\overline{\alpha}_{\mathrm{H}}\sigma^2\|b'\|_{\infty}^2}. \tag{4.15}$$

Then, there exists a constant $\mathfrak{c}_{4.11} > 0$, not depending on N, such that

$$\mathbb{E}((\overline{\theta}_N^{\mathfrak{c},\mathfrak{d}} - \theta_0)^2) \leqslant \frac{\mathfrak{c}_{4.11}}{N}.$$

Moreover, $\overline{\theta}_N^{\mathfrak{c}}$ is a converging estimator of θ_0.

Proof First, $\widehat{\theta}_N = I_N + \Theta_N(\theta_0 - I_N)$ and, on the event Δ_N, $R_N = \overline{\theta}_N - I_N$ is the unique fixed point of the $\mathfrak{c}$-contraction Θ_N by Proposition 4.10. Then, since b' is bounded,

$$\begin{aligned}|\mathbb{E}(\mathfrak{b}(X_T)) - \mathfrak{b}(x_0)| &\leqslant |b(x_0)|\mathbb{E}(|X_T - x_0|) + \frac{\|b'\|_\infty}{2}\mathbb{E}((X_T - x_0)^2)\\ &\leqslant \mathfrak{c}_1(\mathbb{E}((X_T - x_0)^2)^{\frac{1}{2}} + \mathbb{E}((X_T - x_0)^2))\\ &\qquad\qquad \text{with} \quad \mathfrak{c}_1 = |b(x_0)| \vee \frac{\|b'\|_\infty}{2}.\end{aligned}$$

Moreover,

$$\begin{aligned}\mathbb{E}((X_T - x_0)^2) &= \mathbb{E}\left[\left(\theta_0 \int_0^T b(X_s)ds + \sigma B_T\right)^2\right]\\ &\leqslant 2\theta_0^2 T \int_0^T \mathbb{E}(b(X_s)^2)ds + 2\sigma^2\mathbb{E}(B_T^2) = 2\theta_0^2 T^2 \|b\|_f^2 + 2\sigma^2 T^{2\mathrm{H}}.\end{aligned}$$

Then,

$$\begin{aligned}&\exp\left(\frac{\|b'\|_\infty}{\|b\|_f^2}|\mathbb{E}(\mathfrak{b}(X_T)) - \mathfrak{b}(x_0)|\right)\\ &\qquad \leqslant \exp\left(\frac{\mathfrak{c}_1\|b'\|_\infty}{\|b\|_f^2}(\sqrt{2}\theta_0 T\|b\|_f + \sqrt{2}\sigma T^{\mathrm{H}} + 2\theta_0^2 T^2\|b\|_f^2 + 2\sigma^2 T^{2\mathrm{H}})\right)\\ &\qquad = \exp\left(\mathfrak{c}_1\|b'\|_\infty\left(2\theta_0^2 T^2 + \frac{\sqrt{2}\theta_0 T}{\|b\|_f} + \frac{\sigma T^{\mathrm{H}}}{\|b\|_f^2}(\sqrt{2} + 2\sigma T^{\mathrm{H}})\right)\right).\end{aligned}$$

Therefore, by assuming that there exists a known constant $\theta_{\max} > 0$ such that $\theta_0 \in (0, \theta_{\max}]$, T fulfills (4.15) when

$$\begin{aligned}T < \Bigg(\frac{\mathfrak{c}\mathfrak{d}_{\max}}{\overline{\alpha}_{\mathrm{H}}\sigma^2\|b'\|_\infty^2} \exp\Bigg(-\mathfrak{c}_1\|b'\|_\infty\Bigg(2\theta_{\max}^2 T_{\max}^2 + \frac{\theta_{\max}T_{\max}}{\sqrt{\mathfrak{d}_{\max}}}\\ + \frac{\sigma T_{\max}^{\mathrm{H}}}{\mathfrak{d}_{\max}}(1 + \sigma T_{\max}^{\mathrm{H}})\Bigg)\Bigg)\Bigg)^{\frac{1}{2\mathrm{H}}}.\end{aligned}$$

Now, the following proposition provides an asymptotic confidence interval for the computable estimator $\overline{\theta}_N^{\mathfrak{c}}$.

Proposition 4.12 *Assume that $b'(\cdot) \leqslant 0$, $\theta_0 > 0$ and that T satisfies (4.15) with $\mathfrak{c} = 1/2$. Then, for every $\alpha \in (0, 1)$,*

$$\lim_{N\to\infty} \mathbb{P}\left(\theta_0 \in \left[\overline{\theta}_N^{\mathfrak{c}} - \frac{2}{\sqrt{N}D_N}Y_N^{\frac{1}{2}}\mathrm{u}_{1-\frac{\alpha}{4}}; \overline{\theta}_N^{\mathfrak{c}} + \frac{2}{\sqrt{N}D_N}Y_N^{\frac{1}{2}}\mathrm{u}_{1-\frac{\alpha}{4}}\right]\right) \geqslant 1 - \alpha,$$

where $\mathrm{u}_. = \phi^{-1}(\cdot)$, ϕ is the standard normal distribution function, and

$$Y_N = \frac{\sigma^2}{NT^2}\sum_{i=1}^{N}\left(\alpha_{\mathrm{H}}\int_{[0,T]^2}|b(X_s^i)|\cdot|b(X_t^i)|\cdot|t-s|^{2\mathrm{H}-2}dsdt\right.$$
$$\left.+\alpha_{\mathrm{H}}^2\sigma^2\int_{[0,T]^2}\int_0^v\int_0^u|u-\overline{u}|^{2\mathrm{H}-2}|v-\overline{v}|^{2\mathrm{H}-2}b'(X_v^i)b'(X_u^i)d\overline{u}d\overline{v}dudv\right).$$

Proof First, consider

$$U_N = \frac{1}{N}\sum_{i=1}^{N}Z^i,$$

where $Z^1,\dots,Z^N$ are defined by

$$Z^i = \frac{\sigma}{T}\int_0^T b(X_s^i)\delta B_s^i;\ \forall i\in\{1,\dots,N\}.$$

On the one hand, by the (usual) law of large numbers,

$$D_N = \frac{1}{NT}\sum_{i=1}^{N}\int_0^T b(X_s^i)^2ds$$
$$\xrightarrow[N\to\infty]{\mathbb{P}}\mathbb{E}\left(\frac{1}{T}\int_0^T b(X_s)^2ds\right) = \|b\|_f^2 > 0,$$

and by the (usual) central limit theorem,

$$\sqrt{N}U_N \xrightarrow[N\to\infty]{\mathcal{D}} \mathcal{N}(0,\ \mathrm{var}(Z))\quad\text{with}\quad Z = \frac{\sigma}{T}\int_0^T b(X_s)\delta B_s.$$

Then, since $dX_t^i = \theta_0 b(X_t^i)dt + \sigma dB_t^i$ for every $i\in\{1,\dots,N\}$, and by Slutsky's lemma,

$$\sqrt{N}(\widehat{\theta}_N-\theta_0) = \sqrt{N}\frac{U_N}{D_N} \xrightarrow[N\to\infty]{\mathcal{D}} \mathcal{N}\left(0,\ \frac{\mathrm{var}(Z)}{\|b\|_f^4}\right).$$

On the other hand, consider

$$Y_N^* = \frac{\sigma^2}{NT^2}\sum_{i=1}^{N}\left(\alpha_{\mathrm{H}}\int_0^T\int_0^T b(X_s^i)b(X_t^i)|t-s|^{2\mathrm{H}-2}dsdt + R_i\right)$$

where, for every $i\in\{1,\dots,N\}$,

$$\leqslant \frac{\alpha_{\mathrm{H}}\sigma^2}{TD_N}\|b'\|_{\infty}M_N\int_0^T\int_0^t |t-s|^{2\mathrm{H}-2}dsdt \leqslant \frac{\sigma^2\|b'\|_{\infty}}{2T}T^{2\mathrm{H}}\frac{M_N}{D_N} \leqslant \mathfrak{c}_1$$

with

$$\mathfrak{c}_1 = \frac{\mathfrak{c}}{2T\overline{\alpha}_{\mathrm{H}}\|b'\|_{\infty}}.$$

Now, consider $n \in \mathbb{N}^*$. Thanks to a well-known consequence of Picard's fixed point theorem, for every $r \in \mathbb{R}_+$,

$$|(\underbrace{\Theta_N \circ \cdots \circ \Theta_N}_{n} \text{ times})(r) - R_N| \leqslant \frac{\mathfrak{c}^n}{1-\mathfrak{c}}|\Theta_N(r) - r| \quad \text{on} \quad \Delta_N.$$

Then,

$$\begin{aligned}|R_{N,n} - R_N|\mathbf{1}_{\Delta_N} &= |(\Theta_N \circ \cdots \circ \Theta_N)(R_{N,0}) - R_N|\mathbf{1}_{\Delta_N}\\ &\leqslant \frac{\mathfrak{c}^n}{1-\mathfrak{c}}|\Theta_N(0)|\mathbf{1}_{\Delta_N} \leqslant \mathfrak{c}_2\mathfrak{c}^n \quad \text{with} \quad \mathfrak{c}_2 = \frac{\mathfrak{c}_1}{1-\mathfrak{c}}.\end{aligned}$$

So, since T satisfies (4.15), and by (the first part of) Proposition 4.11,

$$\begin{aligned}\mathbb{E}((\overline{\theta}_{N,\psi(N)}^{\mathfrak{c},\mathfrak{d}} - \theta_0)^2) &\leqslant 2\mathbb{E}((R_{N,\psi(N)} - R_N)^2\mathbf{1}_{\{D_N \geqslant \mathfrak{d}\}\cap\Delta_N}) + 2\mathbb{E}((\overline{\theta}_N^{\mathfrak{c},\mathfrak{d}} - \theta_0)^2)\\ &\leqslant \frac{2(1+\mathfrak{c}_{4.11})}{N}\end{aligned} \tag{4.17}$$

Finally, for every $\varepsilon > 0$, since $\overline{\theta}_{N,\psi(N)}^{\mathfrak{c}} = \overline{\theta}_{N,\psi(N)}^{\mathfrak{c},\mathfrak{d}}$ on $\{D_N \geqslant \mathfrak{d}\}$, and by Inequalities (4.13) and (4.17),

$$\begin{aligned}\mathbb{P}(|\overline{\theta}_{N,\psi(N)}^{\mathfrak{c}} - \theta_0| > \varepsilon) &= \mathbb{P}(\{|\overline{\theta}_{N,\psi(N)}^{\mathfrak{c}} - \theta_0| > \varepsilon\} \cap \{D_N < \mathfrak{d}\})\\ &\quad + \mathbb{P}(\{|\overline{\theta}_{N,\psi(N)}^{\mathfrak{c},\mathfrak{d}} - \theta_0| > \varepsilon\} \cap \{D_N \geqslant \mathfrak{d}\})\\ &\leqslant \mathbb{P}(D_N < \mathfrak{d}) + \mathbb{P}(|\overline{\theta}_{N,\psi(N)}^{\mathfrak{c},\mathfrak{d}} - \theta_0| > \varepsilon)\\ &\leqslant \frac{\|b^2\|_f^2}{\mathfrak{d}^2 N} + \frac{1}{\varepsilon^2}\mathbb{E}((\overline{\theta}_{N,\psi(N)}^{\mathfrak{c},\mathfrak{d}} - \theta_0)^2) \leqslant \frac{\mathfrak{c}_3}{N} \xrightarrow[N\to\infty]{} 0\\ &\qquad \text{with} \quad \mathfrak{c}_3 = \frac{\|b^2\|_f^2}{\mathfrak{d}^2} + \frac{2(1+\mathfrak{c}_{4.11})}{\varepsilon^2}.\end{aligned}$$

□

4.2.2.3 Basic Numerical Experiments

In this section, some numerical experiments on the computable approximation of the least squares estimator of θ_0 are presented for the two following models:

(A) $dX_t = \theta_0 b_1(X_t)dt + dB_t$ with $\theta_0 = 1$ and $b_1(x) = -x$,
(B) and $dX_t = \theta_0 b_2(X_t)dt + 0.25dB_t$ with $\theta_0 = 1$ and $b_2(x) = \pi + \arctan(-x)$.

For each model, with $\mathtt{H} = 0.7$ and $\mathtt{H} = 0.9$, $\overline{\theta}_N$ is computed from $N = 1, \dots, 50$ paths of the process X. This experiment is repeated 100 times. The means and the standard deviations of the error $|\overline{\theta}_{50} - \theta_0|$ are stored in Table 4.1. The mean error of $\overline{\theta}_{50}$ is small in the four situations: lower than $4.9 \cdot 10^{-2}$. Since the error standard deviation of $\overline{\theta}_{50}$ is also small in each situations ($< 3.4 \cdot 10^{-2}$), on average, the error of $\overline{\theta}_{50}$ for one repetition of the experiment should be near of its mean error. Note also that for both Model (A) and Model (B), the mean error of $\overline{\theta}_{50}$ is higher when $\mathtt{H} = 0.7$ than when $\mathtt{H} = 0.9$; probably because $\mathtt{H}$ controls the Hölder exponent of the paths of the fractional Brownian motion.

In Fig. 4.3 (resp. Figure 4.4), for $N = 1, \dots, 50$, $\overline{\theta}_N$ and the bounds of the 95%-asymptotic confidence interval (ACI) in Proposition 4.12 are plotted for one of the 100 datasets generated from Model (A) (resp. Model (B)). These figures illustrate both that $\overline{\theta}_N$ is consistent and its rate of convergence.

Table 4.1 Means and standard deviations of the error of $\overline{\theta}_{50}$ (100 repetitions)

	H	Mean error	Error StD.
Model (A)	0.7	0.0489	0.0336
	0.9	0.0186	0.0139
Model (B)	0.7	0.0387	0.0172
	0.9	0.0138	0.0155

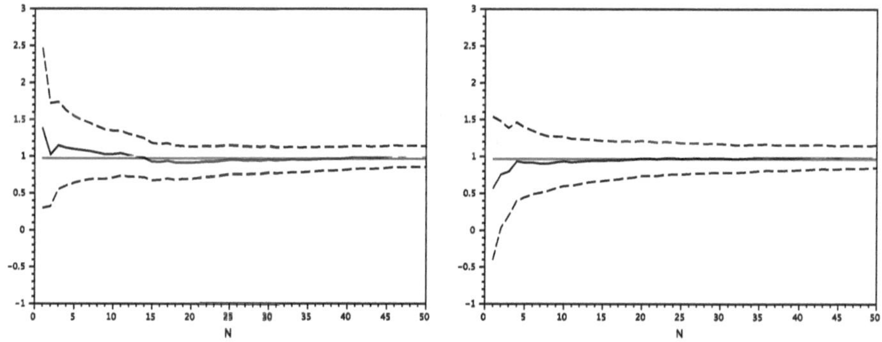

Fig. 4.3 Plots of $N \mapsto \overline{\theta}_N$ (black line) and of the bounds of the 95%-ACIs (dashed black lines) for Model (A) with $\mathtt{H} = 0.7$ (left) and $\mathtt{H} = 0.9$ (right)

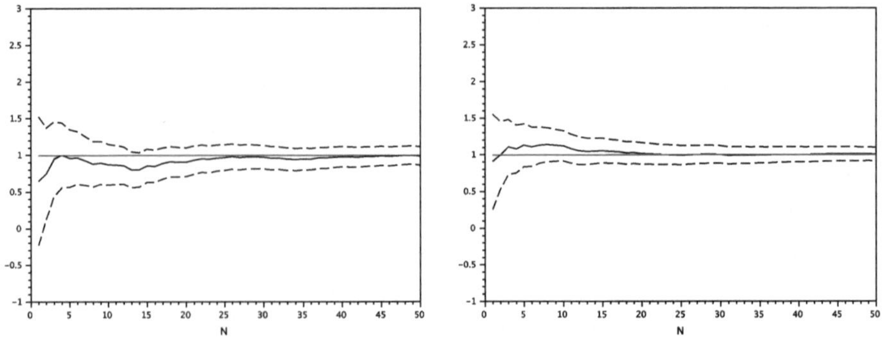

Fig. 4.4 Plots of $N \mapsto \overline{\theta}_N$ (black line) and of the bounds of the 95%-ACIs (dashed black lines) for Model (B) with $\mathrm{H} = 0.7$ (left) and $\mathrm{H} = 0.9$ (right)

4.2.3 The Projection Least Squares Estimator of the Drift Function for Fractional Diffusions

As for the estimation of b_0 in Eq. (4.1) (see Sect. 4.1), except those depending on the empirical process ν_N, all the operators and sets introduced in Chap. 3 remain defined in the same way by replacing the copies of the solution of Eq. (1.1) by $X^1, \dots, X^N$, where

$$X^i = \mathcal{I}(x_0, B^i); \forall i \in \{1, \dots, N\},$$

$B^1, \dots, B^N$ are N independent copies of B, and $\mathcal{I}$ is the solution map for Eq. (4.8). All the integrals with respect to $X^1, \dots, X^N$ or $B^1, \dots, B^N$ are taken in the sense of Skorokhod (see Sect. 4.2.1.2). The projection least squares estimator of b_0 is still defined as a minimizer $\widehat{b}_m$ in $\mathcal{S}_m$ of the objective function γ_N, and the associated truncated estimator $\widetilde{b}_m$ by

$$\widetilde{b}_m(x) = \widehat{b}_m(x)\mathbf{1}_{\Lambda_m}; x \in I.$$

The purpose of this section is to establish a suitable risk bound on $\widetilde{b}_m$ for m fixed, and then to provide a computable approximation of $\widehat{b}_m$ defined as a fixed point in $\mathcal{S}_m$ of a well-chosen random functional (as in Sect. 4.2.2.2). First, note that by Li et al. [16], Theorem 1.3, the density function f remains well-defined, bounded and such that

$$|b_0|^\kappa \in \mathbb{L}^2(\mathbb{R}, f(x)dx); \forall \kappa > 0.$$

Now, consider the empirical process ν_N defined by

$$\nu_N(\tau) := \frac{\sigma}{NT} \sum_{i=1}^{N} \int_0^T \tau(X_s^i)\delta B_s^i.$$

Note that Equality (3.8) remains true again:

$$[\widehat{Z}_m]_j = \langle b_0, \varphi_j \rangle_N + \nu_N(\varphi_j);\ \forall j \in \{1, \dots, m\}.$$

The following theorem, which is the main result of this section, provides a risk bound on $\widetilde{b}_m$.

Theorem 4.4 *If $\varphi_1, \dots, \varphi_m \in C^1(I)$ with $\varphi_1', \dots, \varphi_m'$ bounded on I, and if the matrix Ψ_m is invertible, then there exists a constant $\mathfrak{c}_{4.4,1} > 0$, not depending on m and N, such that*

$$\mathbb{E}(\|\widetilde{b}_m - b_0^I\|_N^2) \leqslant \min_{\tau \in \mathcal{S}_m} \|\tau - b_0^I\|_f^2 + \mathfrak{c}_{4.4,1}\Big[\frac{\|\Psi_m^{-1}\|_{\mathrm{op}}(\mathfrak{L}(m) + \mathfrak{R}(m))}{N} + m\left(1 + \frac{\mathfrak{R}(m)}{\mathfrak{L}(m)}\right)\mathbb{P}(\Omega_m^c)^{\frac{1}{2}} + \mathbb{P}(\Lambda_m^c)^{\frac{1}{2}}\Big],$$

where

$$\mathfrak{R}(m) = 1 \vee \left(\sup_{x \in I} \sum_{j=1}^{m} \varphi_j'(x)^2\right).$$

Under Assumption 3.1 with p large enough to get $\mathfrak{R}(m)\mathfrak{L}(m)^{-1} \lesssim N^{p/2-1}$, there exists a constant $\mathfrak{c}_{4.4,2} > 0$, not depending on m and N, such that

$$\mathbb{E}(\|\widetilde{b}_m - b_0^I\|_N^2) \leqslant \min_{\tau \in \mathcal{S}_m} \|\tau - b_0^I\|_f^2 + \mathfrak{c}_{4.4,2}\frac{\|\Psi_m^{-1}\|_{\mathrm{op}}(\mathfrak{L}(m) + \mathfrak{R}(m))}{N}. \tag{4.18}$$

Proof Since the definition of the empirical projection $\widehat{\Pi}_m$ remains the same as in Chap. 3, as in the proof of Theorem 3.1,

$$\mathbb{E}(\|\widetilde{b}_m - b_0^I\|_N^2) \leqslant \min_{\tau \in \mathcal{S}_m} \mathbb{E}(\|\tau - b_0^I\|_N^2) + \mathbb{E}(\|\widehat{b}_m - \widehat{\Pi}_m(b_0^I)\|_N^2 \mathbf{1}_{\Lambda_m}) + \mathfrak{c}_1\mathbb{P}(\Lambda_m^c)^{\frac{1}{2}}$$

with

$$\mathfrak{c}_1 = \left(\int_I b_0(x)^4 f(x)dx\right)^{\frac{1}{2}}.$$

Moreover, by (3.8), and since Lemma 3.4 has been established for copies of any piecewise continuous stochastic process, as established in the proof of Theorem 3.1 (see the first step),

$$\mathbb{E}(\|\widehat{b}_m - \widehat{\Pi}_m(b_0^I)\|_N^2 \mathbf{1}_{\Lambda_m}) \leqslant \frac{2}{NT}\mathrm{trace}(\Psi_m^{-1}\Phi_{m,\sigma}) + \mathfrak{c}_\Lambda \frac{NT}{\log(NT)\mathfrak{L}(m)}\mathbb{E}\left[\left(\sum_{j=1}^{m} \nu_N(\varphi_j)^2\right)^2\right]^{\frac{1}{2}} \mathbb{P}(\Omega_m^c)^{\frac{1}{2}}$$

with

$$\Phi_{m,\sigma} = NT\mathbb{E}(\widehat{\Delta}_m\widehat{\Delta}_m^*) \quad \text{and} \quad \widehat{\Delta}_m = (\nu_N(\varphi_j))_{j\in\{1,\dots,m\}}.$$

Since the empirical process is not the same as in Chap. 3, suitable bounds on

$$\mathbb{E}\left[\left(\sum_{j=1}^{m}\nu_N(\varphi_j)^2\right)^2\right] \quad \text{and} \quad \text{trace}(\Psi_m^{-1}\Phi_{m,\sigma})$$

need to be established (see Steps 1 and 2) in order to conclude (see Step 3).

Step 1. By Jensen's and Cauchy-Schwarz's inequalities, and since $B^1, \dots, B^N$ are i.i.d. centered processes,

$$\begin{aligned}
\mathbb{E}\left[\left(\sum_{j=1}^{m}\nu_N(\varphi_j)^2\right)^2\right] &\leqslant m\sum_{j=1}^{m}\mathbb{E}(\nu_N(\varphi_j)^4)\\
&= \frac{\sigma^4 m}{N^4T^4}\sum_{j=1}^{m}\left[\sum_{i=1}^{N}\mathbb{E}\left[\left(\int_0^T \varphi_j(X_s^i)\delta B_s^i\right)^4\right]\right.\\
&\quad \left.+3\sum_{i\neq k}\mathbb{E}\left[\left(\int_0^T \varphi_j(X_s^i)\delta B_s^i\right)^2\left(\int_0^T \varphi_j(X_s^k)\delta B_s^k\right)^2\right]\right]\\
&\leqslant \frac{4\sigma^4 m}{N^2T^4}\sum_{j=1}^{m}\mathbb{E}(\delta_j^4)
\end{aligned}$$

with

$$\delta_j = \int_0^T \varphi_j(X_s)\delta B_s;\ \forall j \in \{1,\dots,m\}.$$

By Proposition 4.8,

$$\begin{aligned}
\sum_{j=1}^{m}\mathbb{E}(\delta_j^4) &\leqslant \mathfrak{c}_{4.8}\mathfrak{m}_{4,M}\sum_{j=1}^{m}\left[\left(\int_0^T \mathbb{E}(|\varphi_j(X_s)|^{\frac{1}{\mathtt{H}}})ds\right)^{4\ \mathtt{H}} + \left(\int_0^T \mathbb{E}(\varphi_j'(X_s)^4)^{\frac{1}{4\ \mathtt{H}}}ds\right)^{4\ \mathtt{H}}\right]\\
&\leqslant \mathfrak{c}_{4.8}\mathfrak{m}_{4,M}T^{4\ \mathtt{H}}\underbrace{\sum_{j=1}^{m}\int_{-\infty}^{\infty}(\varphi_j(x)^4 + \varphi_j'(x)^4)f(x)dx}_{\leqslant\mathfrak{L}(m)^2+\mathfrak{R}(m)^2}.
\end{aligned}$$

So,

$$\mathbb{E}\left[\left(\sum_{j=1}^{m}\nu_N(\varphi_j)^2\right)^2\right] \leqslant \mathfrak{c}_2\frac{m(\mathfrak{L}(m)^2+\mathfrak{R}(m)^2)}{N^2} \quad\text{with}\quad \mathfrak{c}_2 = \frac{4\sigma^4\mathfrak{c}_{4.8}\mathfrak{m}_{4,M}}{T^{4-4\mathrm{H}}}.$$

Therefore, there exists a constant $\mathfrak{c}_3 > 0$, not depending on m and N, such that

$$\mathfrak{c}_\Lambda\frac{NT}{\log(NT)\mathfrak{L}(m)}\mathbb{E}\left[\left(\sum_{j=1}^{m}\nu_N(\varphi_j)^2\right)^2\right]^{\frac{1}{2}} \leqslant \mathfrak{c}_3 m^{\frac{1}{2}}\left(1+\frac{\mathfrak{R}(m)}{\mathfrak{L}(m)}\right).$$

Step 2. First, let us show that $\Phi_{m,\sigma}$ is a positive semidefinite symmetric matrix. Indeed, for every $y \in \mathbb{R}^m$,

$$y^*\Phi_{m,\sigma}y = \frac{\sigma^2}{NT}\mathbb{E}\left[\left(\sum_{i=1}^{N}\int_0^T\left(\sum_{j=1}^{m}y_j\varphi_j(X_s^i)\right)\delta B_s^i\right)^2\right] \geqslant 0.$$

Then, by Proposition 3.1 (6),

$$\mathrm{trace}(\Psi_m^{-1}\Phi_{m,\sigma}) \leqslant \|\Psi_m^{-1}\|_{\mathrm{op}}\mathrm{trace}(\Phi_{m,\sigma}) \leqslant \frac{\sigma^2\|\Psi_m^{-1}\|_{\mathrm{op}}}{T}\sum_{j=1}^{m}\mathbb{E}(\delta_j^2).$$

By Proposition 4.8,

$$\begin{aligned}\sum_{j=1}^{m}\mathbb{E}(\delta_j^2) &\leqslant \mathfrak{c}_{4.8}\mathfrak{m}_{2,M}\sum_{j=1}^{m}\left[\left(\int_0^T\mathbb{E}(|\varphi_j(X_s)|^{\frac{1}{\mathrm{H}}})ds\right)^{2\mathrm{H}}+\left(\int_0^T\mathbb{E}(\varphi_j'(X_s)^2)^{\frac{1}{2\mathrm{H}}}ds\right)^{2\mathrm{H}}\right]\\ &\leqslant \mathfrak{c}_{4.8}\mathfrak{m}_{2,M}T^{2\mathrm{H}}\underbrace{\sum_{j=1}^{m}\int_{-\infty}^{\infty}(\varphi_j(x)^2+\varphi_j'(x)^2)f(x)dx}_{\leqslant\mathfrak{L}(m)+\mathfrak{R}(m)}.\end{aligned}$$

Therefore,

$$\mathrm{trace}(\Psi_m^{-1}\Phi_{m,\sigma}) \leqslant \mathfrak{c}_4\|\Psi_m^{-1}\|_{\mathrm{op}}(\mathfrak{L}(m)+\mathfrak{R}(m)) \quad\text{with}\quad \mathfrak{c}_4 = \mathfrak{c}_{4.8}\mathfrak{m}_{2,M}\sigma^2T^{2\mathrm{H}-1}.$$

Step 3 (conclusion). By the two previous steps,

$$\begin{aligned}\mathbb{E}(\|\widetilde{b}_m-b_0^I\|_N^2) \leqslant& \min_{\tau\in S_m}\|\tau-b_0^I\|_f^2+\frac{2\mathfrak{c}_4}{NT}\|\Psi_m^{-1}\|_{\mathrm{op}}(\mathfrak{L}(m)+\mathfrak{R}(m))\\ &+\mathfrak{c}_3 m^{\frac{1}{2}}\left(1+\frac{\mathfrak{R}(m)}{\mathfrak{L}(m)}\right)\mathbb{P}(\Omega_m^c)^{\frac{1}{2}}+\mathfrak{c}_1\mathbb{P}(\Lambda_m^c)^{\frac{1}{2}}.\end{aligned}$$

Under Assumption 3.1 with p large enough to get $\mathfrak{R}(m)\mathfrak{L}(m)^{-1} \lesssim N^{p/2-1}$, by Propositions 3.5 and 3.7, there exists a constant $\mathfrak{c}_5 > 0$, not depending on m and N, such that

$$\mathbb{E}(\|\widetilde{b}_m - b_0^I\|_N^2) \leqslant \min_{\tau \in S_m} \|\tau - b_0^I\|_f^2 + \mathfrak{c}_5 \frac{\|\Psi_m^{-1}\|_{\mathrm{op}}(\mathfrak{L}(m) + \mathfrak{R}(m))}{N}.$$

□

Let us make some remarks about Theorem 4.4:

- At least for the usual projection spaces, note that the order of the *variance term* in Inequality (4.18) is degraded with respect to that of Theorems 3.1 and 4.1, because φ_j' is involved in the bound on

$$\mathbb{E}\left[\left(\int_0^T \varphi_j(X_s)\delta X_s\right)^2\right]; \ j \in \{1, \dots, m\}$$

provided in Proposition 4.8. Precisely, the *variance term* in Inequality (4.18) is of order

$$\rho_1(m) = \frac{\|\Psi_m^{-1}\|_{\mathrm{op}}(\mathfrak{L}(m) + \mathfrak{R}(m))}{N} \quad \text{instead of} \quad \frac{m}{N}.$$

In the proof of Theorem 4.4, by applying Proposition 3.1 (6) to $A = \Psi_m^{-1}\Phi_{m,\sigma}$ and $B = I$ (as in the proof of Theorem 3.1) instead of $A = \Phi_{m,\sigma}$ and $B = \Psi_m^{-1}$, one can establish a risk bound on our truncated projection least squares estimator of b_0 with a *variance term* of order

$$\rho_2(m) = \frac{m}{N}(1 + \|\Psi_m^{-1}\|_{\mathrm{op}}\mathfrak{R}(m)).$$

However, for the usual projection spaces, $\rho_1(m) \lesssim \rho_2(m)$.
- Consider $\mathtt{l}, \mathtt{r} \in \mathbb{R}$ satisfying $\mathtt{l} < \mathtt{r}$, and assume that $(\varphi_1, \dots, \varphi_m)$ is the $[\mathtt{l}, \mathtt{r}]$-supported m-dimensional trigonometric basis. As already established,

$$\mathfrak{L}(m) \leqslant \mathfrak{c}_\varphi m \quad \text{and} \quad \|\Psi_m^{-1}\|_{\mathrm{op}} \leqslant \frac{1}{\mathtt{m}}.$$

Moreover, since $\varphi_1' \equiv 0$, and since

$$\varphi_{2j+1}' = \frac{2\pi j}{\mathtt{r} - \mathtt{l}}\varphi_{2j} \quad \text{and} \quad \varphi_{2j}' = -\frac{2\pi j}{\mathtt{r} - \mathtt{l}}\varphi_{2j+1}; \forall j \in \{1, \dots, m\},$$

there exists a constant $\overline{\mathfrak{c}}_\varphi > 0$, not depending on m and N, such that

$$\mathfrak{R}(m) \leqslant \overline{\mathfrak{c}}_\varphi m^3.$$

So, the *variance term* in Inequality (4.18) is of order m^3/N. Now, consider $\beta \in \mathbb{N}^*$, and assume that $(b_0)_{|I} \in \mathbb{W}_0^{\beta}([\mathtt{l}, \mathtt{r}])$. As already mentioned, this leads to

$$\min_{\tau \in \mathcal{S}_m} \|\tau - b_0^I\|_f^2 \leqslant \mathfrak{c}_{\beta,\mathtt{l},\mathtt{r}} m^{-2\beta}.$$

In conclusion, by Theorem 4.4, there exists a constant $\mathfrak{c}_{4.4,2} > 0$, not depending on m and N, such that

$$\mathbb{E}(\|\widetilde{b}_m - b_0^I\|_N^2) \leqslant \mathfrak{c}_{4.2,2}\left(m^{-2\beta} + \frac{m^3}{N}\right),$$

and then the bias-variance tradeoff is reached by (the risk bound on) $\widetilde{b}_m$ for m of order $N^{1/(3+2\beta)}$.

As $\widehat{\theta}_N$ in Sect. 4.2.2, $\widehat{b}_m$ is not directly computable because of Skorokhod's integral. However, one can extend the method of Sect. 4.2.2.2 to $\widehat{b}_m$, again by using the relationship between the Skorokhod integral and the pathwise stochastic integral provided by Proposition 4.7. Assume that $\varphi_1, \dots, \varphi_m \in C^1(I)$ with $\varphi_1', \dots, \varphi_m'$ bounded on I, and let I_m be the random function defined on $\mathbb{R}$ by

$$I_m(x) = \sum_{j=1}^{m}\left[\widehat{\Psi}_m^{-1}\left(\frac{1}{NT}\sum_{i=1}^{N}\int_0^T \varphi_\ell(X_s^i)dX_s^i\right)_\ell\right]_j \varphi_j(x);\ \forall x \in \mathbb{R}.$$

Moreover, for every I-supported function $\rho : \mathbb{R} \to \mathbb{R}$ such that $\rho \in C^1(I)$, consider

$$\Theta_m(\rho) = \sum_{j=1}^{m}[\widehat{\Psi}_m^{-1}\theta_m(\rho)]_j\varphi_j$$

with

$$\theta_m(\rho) = \left(-\frac{\alpha_{\mathrm{H}}\sigma^2}{NT}\sum_{i=1}^{N}\int_0^T\int_0^t \varphi_j'(X_t^i)\exp\left[\int_s^t(\rho + I_m)'(X_u^i)du\right]|t-s|^{2\mathrm{H}-2}dsdt\right)_j.$$

By Proposition 4.7, and since $\widehat{b}_m$ is a converging estimator of b_0 by Theorem 4.4,

$$\widehat{b}_m - I_m = \Theta_m(b_0 - I_m) \approx \Theta_m(\widehat{b}_m - I_m).$$

This legitimates to consider the estimator $\overline{b}_m = I_m + R_m$ of b_0, where R_m is a fixed point of the map Θ_m in $\mathcal{S}_m$. For numerical purposes, let us also consider the estimator $\overline{b}_{m,n} = I_m + R_{m,n}$, where the sequence $(R_{m,n})_{n\in\mathbb{N}}$ is defined by $R_{m,0} \in \mathcal{S}_m$ and

$$R_{m,n+1} = \Theta_m(R_{m,n});\ n \in \mathbb{N}.$$

4.2.4 A Projection Least Squares Estimator of the Drift Function for Some Non-autonomous Fractional Diffusions

Let us consider the stochastic process $Z = (Z_t)_{t\in[0,T]}$ defined by

$$Z_t = \int_0^t J_0(s)d\langle M\rangle_s + M_t;\ \forall t \in [0, T] \tag{4.19}$$

where $M = (M_t)_{t\in[0,T]} \neq 0$ is a continuous and square integrable martingale vanishing at 0, and J_0 is an unknown function which almost surely belongs to $\mathbb{L}^2([0, T], d\langle M\rangle_t)$. Under these conditions on M and J_0, the quadratic variation $\langle M\rangle_t$ of M is well-defined for any $t \in [0, T]$, and the Riemann-Stieljès integral of J_0 with respect to $s \mapsto \langle M\rangle_s$ on $[0, t]$ exists and is finite. By assuming that $s \mapsto \langle M\rangle_s$ is a deterministic function, this section deals first with risk bounds and model selection for a copies-based projection least squares estimator of J_0. Now, let $X = (X_t)_{t\in[0,T]}$ be the non-autonomous fractional diffusion defined by

$$X_t = X_0 + \int_0^t V(X_s)(b_0(s)ds + \sigma(s)dB_s);\ t \in [0, T] \tag{4.20}$$

where X_0 is a $\mathbb{R}^*$-valued random variable, $B = (B_t)_{t\in[0,T]}$ is a fractional Brownian motion of Hurst parameter $\mathtt{H} \in [1/2, 1)$, the stochastic integral with respect to B is taken pathwise when $\mathtt{H} > 1/2$ and in the sense of Itô when $\mathtt{H} = 1/2$, and $V : \mathbb{R} \to \mathbb{R}$, $\sigma : [0, T] \to \mathbb{R}^*$ and $b_0 : [0, T] \to \mathbb{R}$ are at least continuous functions. An *appropriate* transformation allows to rewrite Eq. (4.20) as a model of type (4.19) driven by the Molchan martingale, which quadratic variation is $t^{2-2\mathtt{H}}$ for every $t \in [0, T]$. So, this section also deals with a risk bound on an estimator of b_0/σ derived from the aforementioned projection least squares estimator of J_0. Finally, some applications in mathematical finance are provided.

4.2.4.1 A Projection Least Squares Estimator of J_0

In the sequel, the quadratic variation $\langle M\rangle = (\langle M\rangle_t)_{t\in[0,T]}$ of M fulfills the following assumption.

Assumption 4.2 The (nonnegative, increasing and continuous) process $\langle M\rangle$ is a deterministic function.

Assumption 4.2 is fulfilled by the Brownian motion and, more generally, by any martingale $(M_t)_{t\in[0,T]}$ such that

$$M_t = \int_0^t \zeta(s)dW_s;\ \forall t \in [0, T],$$

where W is a Brownian motion and $\zeta \in \mathbb{L}^2([0,T],dt)$. For some results, $\langle M\rangle$ fulfills the following stronger assumption.

Assumption 4.3 There exists $\mu \in C^0((0,T];\mathbb{R}_+)$ such that $\mu(\cdot)^{-1}$ is continuous from $[0,T]$ into $\mathbb{R}_+$, and such that

$$\langle M\rangle_t = \int_0^t \mu(s)ds;\ \forall t \in [0,T].$$

Here again, Assumption 4.3 is fulfilled by the Brownian motion. Assumption 4.3 is also fulfilled by any martingale $(M_t)_{t\in[0,T]}$ such that

$$M_t = \int_0^t \zeta(s)dW_s;\ \forall t \in [0,T],$$

where W is a Brownian motion, $\zeta \in C^0((0,T])$ and $\zeta(\cdot)^{-1}$ is continuous from $[0,T]$ into $\mathbb{R}$. This last condition is satisfied, for instance, when ζ is a (c,∞)-valued function with $c>0$, or when $\zeta(t)=t^{-\kappa}$ ($\kappa>0$) for every $t\in(0,T]$. For instance, let M be the Molchan martingale, which is defined by

$$M_t = \int_0^t \ell(t,s)dB_s;\ \forall t \in [0,T],$$

where B is a fractional Brownian motion of Hurst parameter $\mathrm{H} \in [1/2,1)$, and

$$\ell(t,s) = \mathfrak{c}_{\mathrm{H}} s^{\frac{1}{2}-\mathrm{H}}(t-s)^{\frac{1}{2}-\mathrm{H}}\mathbf{1}_{(0,t)}(s);\ \forall s,t \in [0,T]$$

with

$$\mathfrak{c}_{\mathrm{H}} = \left(\frac{\Gamma(3-\mathrm{H})}{2\mathrm{H}\Gamma(3/2-\mathrm{H})^3\Gamma(\mathrm{H}+1/2)}\right)^{\frac{1}{2}}.$$

Since

$$M_t = (2-2\mathrm{H})^{\frac{1}{2}}\int_0^t s^{\frac{1}{2}-\mathrm{H}}dW_s;\ \forall t \in [0,T],$$

where W is the Brownian motion driving the Mandelbrot-Van Ness representation of the fractional Brownian motion B, the Molchan martingale fulfills Assumption 4.3 with $\mu(t) = (2-2\mathrm{H})t^{1-2\mathrm{H}}$ for every $t \in (0,T]$.

In order to define our projection least squares estimator of J_0, let us consider N independent copies $M^1,\dots,M^N$ of M, the copies $Z^1,\dots,Z^N$ of Z defined by

$$Z^i_\cdot = \int_0^\cdot J_0(s)d\langle M^i\rangle_s + M^i;\ \forall i \in \{1,\dots,N\},$$

and the objective function γ_N defined by

$$\gamma_N(J) = \frac{1}{N}\sum_{i=1}^{N}\left(\int_0^T J(s)^2 d\langle M^i\rangle_s - 2\int_0^T J(s)dZ_s^i\right) ; \forall J \in \mathcal{S}_m,$$

where $m \in \{1, \dots, N\}$, $\mathcal{S}_m = \text{span}\{\varphi_1, \dots, \varphi_m\}$ and $(\varphi_1, \dots, \varphi_m)$ is an orthonormal family of $\mathbb{L}^2([0, T], dt)$. Note that since $t \in [0, T] \mapsto \langle M\rangle_t$ is nonnegative, increasing and continuous, and since the φ_j's are continuous from $[0, T]$ into $\mathbb{R}$, the objective function γ_N is well-defined. Let us show that γ_N has a unique minimizer in $\mathcal{S}_m$. For $J = \sum_{j=1}^m \theta_j\varphi_j$ with $\theta_1, \dots, \theta_m \in \mathbb{R}$,

$$\begin{aligned}\nabla\gamma_N(J) &= \left(\frac{2}{N}\sum_{i=1}^{N}\left[\int_0^T \varphi_j(s)J(s)d\langle M^i\rangle_s - \int_0^T \varphi_j(s)dZ_s^i\right]\right)_{j\in\{1,\dots,m\}}\\ &= 2(\Psi_m\theta - \widehat{Z}_m),\end{aligned}$$

where $\Psi_m = (\langle\varphi_j, \varphi_\ell\rangle_{\langle M\rangle})_{j,\ell}$, $\langle\cdot,\cdot\rangle_{\langle M\rangle}$ is the usual inner product on $\mathbb{L}^2([0, T], d\langle M\rangle_t)$, and

$$\widehat{Z}_m = \left(\frac{1}{N}\sum_{i=1}^{N}\int_0^T \varphi_j(s)dZ_s^i\right)_{j\in\{1,\dots,m\}}.$$

The symmetric matrix Ψ_m is positive semidefinite because under Assumption 4.2, for every $x \in \mathbb{R}^m$,

$$x^*\Psi_m x = \int_0^T \left(\sum_{j=1}^m x_j\varphi_j(s)\right)^2 d\langle M\rangle_s \geqslant 0.$$

In fact, since $\varphi_1, \dots, \varphi_m$ are linearly independent, Ψ_m is a positive definite matrix, and then

$$\widehat{J}_m = \sum_{j=1}^m \widehat{\theta}_j\varphi_j \quad \text{with} \quad \widehat{\theta} = \Psi_m^{-1}\widehat{Z}_m$$

is the only minimizer of γ_N in $\mathcal{S}_m$ called the projection least squares estimator of J_0. In practice, since the process Z cannot be observed continuously on the time interval $[0, T]$, the vector $\widehat{Z}_m$ must be replaced by the approximation

$$\widehat{Z}_{m,n} = \left(\frac{1}{N}\sum_{i=1}^{N}\sum_{k=0}^{n-1}\varphi_j(t_k)(Z_{t_{k+1}}^i - Z_{t_k}^i)\right)_{j\in\{1,\dots,m\}}$$

in the definition of $\widehat{J}_m$, where $t_k = kT/n$ for every $k \in \{0, \dots, n\}$. This leads to the discrete-time estimator

$$\widehat{J}_{m,n} = \sum_{j=1}^{m} [\widehat{\theta}_n]_j \varphi_j \quad \text{with} \quad \widehat{\theta}_n = \Psi_m^{-1} \widehat{Z}_{m,n}.$$

4.2.4.2 Risk Bounds and Model Selection

First, let us introduce the empirical process ν_N defined by

$$\nu_N(\tau) := \frac{1}{N} \sum_{i=1}^{N} \int_0^T \tau(s) dM_s^i.$$

The following proposition provides a risk bound on $\widehat{J}_m$ for m fixed.

Proposition 4.14 *Under Assumption 4.2,*

$$\mathbb{E}(\|\widehat{J}_m - J_0\|_{\langle M \rangle}^2) \leqslant \min_{J \in \mathcal{S}_m} \|J - J_0\|_{\langle M \rangle}^2 + \frac{2m}{N}. \tag{4.21}$$

Proof For every $J, \overline{J} \in \mathcal{S}_m$,

$$\begin{aligned} \gamma_N(J) - \gamma_N(\overline{J}) &= \|J\|_{\langle M \rangle}^2 - \|\overline{J}\|_{\langle M \rangle}^2 - \frac{2}{N} \sum_{i=1}^{N} \int_0^T (J(s) - \overline{J}(s)) dZ_s^i \\ &= \|J - J_0\|_{\langle M \rangle}^2 - \|\overline{J} - J_0\|_{\langle M \rangle}^2 - \frac{2}{N} \sum_{i=1}^{N} \int_0^T (J(s) - \overline{J}(s)) dM_s^i. \end{aligned}$$

Moreover,

$$\gamma_N(\widehat{J}_m) \leqslant \gamma_N(J); \forall J \in \mathcal{S}_m.$$

So, for every $J \in \mathcal{S}_m$,

$$\|\widehat{J}_m - J_0\|_{\langle M \rangle}^2 \leqslant \|J - J_0\|_{\langle M \rangle}^2 + \frac{2}{N} \sum_{i=1}^{N} \int_0^T (\widehat{J}_m(s) - J(s)) dM_s^i$$

and then,

$$\mathbb{E}(\|\widehat{J}_m - J_0\|_{\langle M \rangle}^2) \leqslant \|J - J_0\|_{\langle M \rangle}^2 + 2\mathbb{E}\left(\frac{1}{N} \sum_{i=1}^{N} \int_0^T \widehat{J}_m(s) dM_s^i\right).$$

Consider

$$\mathfrak{I}_0 = (\langle \varphi_j, J_0 \rangle_{\langle M \rangle})_{j \in \{1,\dots,m\}} \quad \text{and} \quad \widehat{\Delta}_m = (\nu_N(\varphi_j))_{j \in \{1,\dots,m\}}.$$

Since $\widehat{\Delta}_m$ is a centered random vector,

$$\mathbb{E}\left(\frac{1}{N}\sum_{i=1}^{N}\int_0^T \widehat{J}_m(s)dM_s^i\right) = \mathbb{E}(\langle\widehat{\theta}, \widehat{\Delta}_m\rangle_{2,m})$$

$$= \mathbb{E}(\widehat{\Delta}_m^*\Psi_m^{-1}(\mathfrak{j}_0 + \widehat{\Delta}_m)) = \mathbb{E}(\widehat{\Delta}_m^*\Psi_m^{-1}\widehat{\Delta}_m).$$

Moreover, since $M^1, \dots, M^N$ are independent copies of M, and since Ψ_m is a symmetric matrix,

$$\begin{aligned}
\mathbb{E}(\widehat{\Delta}_m^*\Psi_m^{-1}\widehat{\Delta}_m) &= \sum_{j,\ell=1}^{m}\mathbb{E}([\widehat{\Delta}_m]_j[\widehat{\Delta}_m]_\ell)[\Psi_m^{-1}]_{j,\ell}\\
&= \frac{1}{N}\sum_{j,\ell=1}^{m}[\Psi_m^{-1}]_{j,\ell}\int_0^T \varphi_j(s)\varphi_\ell(s)d\langle M\rangle_s\\
&= \frac{1}{N}\sum_{\ell=1}^{m}\sum_{j=1}^{m}[\Psi_m]_{\ell,j}[\Psi_m^{-1}]_{j,\ell}\\
&= \frac{1}{N}\sum_{\ell=1}^{m}[\Psi_m\Psi_m^{-1}]_{\ell,\ell} = \frac{m}{N}.
\end{aligned}$$

Therefore,

$$\mathbb{E}(\|\widehat{J}_m - J_0\|_{\langle M\rangle}^2) \leqslant \min_{J\in\mathcal{S}_m}\|J - J_0\|_{\langle M\rangle}^2 + \frac{2m}{N}.$$

□

Under Assumption 4.3, the following corollary provides a suitable bias term in Inequality (4.21).

Corollary 4.1 *Under Assumption 4.3,*

$$\mathbb{E}(\|\widehat{J}_m - J_0\|^2) \leqslant \|\mu(\cdot)^{-1}\|_\infty\left(\|\Pi_m^\mu(\mu^{\frac{1}{2}}J_0) - \mu^{\frac{1}{2}}J_0\|^2 + \frac{2m}{N}\right),$$

where Π_m^μ is the orthogonal projection from $\mathbb{L}^2([0,T],dt)$ onto

$$\mathcal{S}_m^\mu = \{\iota \in \mathbb{L}^2([0,T],dt) : \exists\varphi\in\mathcal{S}_m,\ \forall t\in(0,T],\ \iota(t) = \mu(t)^{\frac{1}{2}}\varphi(t)\}.$$

Proof Under Assumption 4.3,

$$\min_{J\in\mathcal{S}_m}\|J - J_0\|_{\langle M\rangle}^2 = \min_{J\in\mathcal{S}_m}\|\mu^{\frac{1}{2}}(J - J_0)\|^2 = \min_{\iota\in\mathcal{S}_m^\mu}\|\iota - \mu^{\frac{1}{2}}J_0\|^2.$$

Since $\mathcal{S}_m^\mu$ is a closed vector subspace of $\mathbb{L}^2([0,T],dt)$,

$$\min_{\iota \in \mathcal{S}_m^\mu} \|\iota - \mu^{\frac{1}{2}} J_0\|^2 = \|\Pi_m^\mu(\mu^{\frac{1}{2}} J_0) - \mu^{\frac{1}{2}} J_0\|^2. \tag{4.22}$$

Moreover, since $\mu(\cdot)^{-1}$ is continuous from $[0, T]$ into $\mathbb{R}_+$ under Assumption 4.3,

$$\begin{aligned}\|\widehat{J}_m - J_0\|^2 &= \|\mu^{-\frac{1}{2}}(\widehat{J}_m - J_0)\|_{\langle M\rangle}^2 \\ &\leqslant \|\mu(\cdot)^{-1}\|_\infty \|\widehat{J}_m - J_0\|_{\langle M\rangle}^2.\end{aligned} \tag{4.23}$$

Equality (4.22) together with Inequality (4.23) allow to conclude.

□

For instance, assume that $\mathcal{S}_m = \text{span}\{\varphi_1^\mu, \dots, \varphi_m^\mu\}$ where, for every $t \in [0, T]$ and $j \in \mathbb{N}^*$ satisfying $2j + 1 \leqslant m$,

$$\begin{aligned}\varphi_1^\mu(t) &= \sqrt{\frac{1}{\mu(t)T}}, \\ \varphi_{2j}^\mu(t) &= \sqrt{\frac{2}{\mu(t)T}} \cos\left(2\pi j \frac{t}{T}\right) \text{ and} \\ \varphi_{2j+1}^\mu(t) &= \sqrt{\frac{2}{\mu(t)T}} \sin\left(2\pi j \frac{t}{T}\right).\end{aligned}$$

Note that $\varphi_1^\mu, \dots, \varphi_m^\mu$ are linearly independent because $(\varphi_1^\mu, \dots, \varphi_m^\mu)$ is an orthonormal family of $\mathbb{L}^2([0, T], d\langle M\rangle_t)$. So, the basis $(\varphi_1, \dots, \varphi_m)$ of $\mathcal{S}_m$, orthonormal in $\mathbb{L}^2([0, T], dt)$, is obtained from $(\varphi_1^\mu, \dots, \varphi_m^\mu)$ via the Gram-Schmidt process. Consider also $\beta \in \mathbb{N}^*$, the Fourier-Sobolev space

$$\mathbb{W}_2^\beta([0, T]) = \left\{\iota : [0, T] \to \mathbb{R}\ \beta\text{times differentiable} : \int_0^T \iota^{(\beta)}(s)^2 ds < \infty\right\},$$

and assume that there exists $\iota_0 \in \mathbb{W}_2^\beta([0, T])$ such that

$$\iota_0(t) = \mu(t)^{\frac{1}{2}} J_0(t);\ \forall t \in (0, T].$$

By DeVore and Lorentz [17], Corollary 2.4, p. 205, there exists a constant $\mathfrak{c}_{\beta,T} > 0$, not depending on m and N, such that

$$\|\Pi_m^\mu(\mu^{\frac{1}{2}} J_0) - \mu^{\frac{1}{2}} J_0\|^2 = \|\Pi_m^\mu(\iota_0) - \iota_0\|^2 \leqslant \mathfrak{c}_{\beta,T} m^{-2\beta}.$$

Therefore, by Corollary 4.1,

$$\mathbb{E}(\|\widehat{J}_m - J_0\|^2) \leqslant \|\mu(\cdot)^{-1}\|_\infty \left(\mathfrak{c}_{\beta,T} m^{-2\beta} + \frac{2\,m}{N}\right).$$

In practice, the dimension m of $\mathcal{S}_m$ needs to be selected from data because the one for which $\widehat{J}_m$ reaches the bias-variance tradeoff depends on some unknown regularity parameters of J_0. Let us consider $m_N \in \{1, \dots, N\}$, $\mathcal{M}_N = \{1, \dots, m_N\}$ and

$$\widehat{m} = \arg\min_{m \in \mathcal{M}_N} \{\gamma_N(\widehat{J}_m) + \text{pen}(m)\} \tag{4.24}$$

where

$$\text{pen}(m) = \mathfrak{c}_{\text{cal}} \frac{m}{N}; \forall m \in \mathcal{M}_N,$$

and the constant $\mathfrak{c}_{\text{cal}} > 0$ needs to be calibrated in practice. In the sequel, the φ_j's fulfill the following assumption.

Assumption 4.4 For every $m, m' \in \{1, \dots, N\}$, if $m > m'$, then $\mathcal{S}_{m'} \subset \mathcal{S}_m$.

The following theorem provides a risk bound on the adaptive estimator $\widehat{J}_{\widehat{m}}$.

Theorem 4.5 *Under Assumptions 4.2 and 4.4, there exists a constant* $\mathfrak{c}_{4.5,1} > 0$*, not depending on* N*, such that*

$$\mathbb{E}(\|\widehat{J}_{\widehat{m}} - J_0\|_{\langle M \rangle}^2) \leqslant \mathfrak{c}_{4.5,1} \left(\min_{m \in \mathcal{M}_N} \{\mathbb{E}(\|\widehat{J}_m - J_0\|_{\langle M \rangle}^2) + \text{pen}(m)\} + \frac{1}{N} \right).$$

Moreover, under Assumption 4.3, there exists a constant $\mathfrak{c}_{4.5,2} > 0$*, not depending on* N*, such that*

$$\mathbb{E}(\|\widehat{J}_{\widehat{m}} - J_0\|^2) \leqslant \mathfrak{c}_{4.5,2} \|\mu(\cdot)^{-1}\|_\infty \times \left(\min_{m \in \mathcal{M}_N} \left\{ \|\Pi_m^\mu(\mu^{\frac{1}{2}} J_0) - \mu^{\frac{1}{2}} J_0\|^2 + \frac{m}{N} \right\} + \frac{1}{N} \right).$$

Proof As established in the proof of Proposition 4.14, for every $J, \overline{J} \in \mathcal{S}_m$,

$$\gamma_N(J) - \gamma_N(\overline{J}) = \|J - J_0\|_{\langle M \rangle}^2 - \|\overline{J} - J_0\|_{\langle M \rangle}^2 - \frac{2}{N} \sum_{i=1}^N \int_0^T (J(s) - \overline{J}(s)) dM_s^i.$$

Moreover, for any $m \in \mathcal{M}_N$,

$$\gamma_N(\widehat{J}_{\widehat{m}}) + \text{pen}(\widehat{m}) \leqslant \gamma_N(\widehat{J}_m) + \text{pen}(m)$$

and then,

$$\gamma_N(\widehat{J}_{\widehat{m}}) - \gamma_N(\widehat{J}_m) \leqslant \text{pen}(m) - \text{pen}(\widehat{m}).$$

So, since $\mathcal{S}_m + \mathcal{S}_{\widehat{m}} \subset \mathcal{S}_{m \vee \widehat{m}}$ under Assumption 4.4, and since $2ab \leqslant a^2 + b^2$ for every $a, b \in \mathbb{R}$,

$$\|\widehat{J}_{\widehat{m}} - J_0\|_{\langle M \rangle}^2 \leqslant \|\widehat{J}_m - J_0\|_{\langle M \rangle}^2 + 2\nu_N(\widehat{J}_{\widehat{m}} - \widehat{J}_m) + \text{pen}(m) - \text{pen}(\widehat{m})$$

$$\leqslant \|\widehat{J}_m - J_0\|_{\langle M\rangle}^2$$
$$+ \frac{1}{4}\|\widehat{J}_{\widehat{m}} - \widehat{J}_m\|_{\langle M\rangle}^2 + 4\left(\left[\sup_{\tau\in\mathcal{B}_{m,\widehat{m}}} |\nu_N(\tau)|\right]^2 - p(m,\widehat{m})\right)_+$$
$$+ \text{pen}(m) + 4p(m,\widehat{m}) - \text{pen}(\widehat{m}),$$

where

$$\mathcal{B}_{m,m'} = \{\varphi \in \mathcal{S}_{m\vee m'} : \|\varphi\|_{\langle M\rangle} = 1\}$$
$$\text{and} \quad p(m,m') = \frac{\mathfrak{c}_{\text{cal}}}{4}\cdot\frac{m\vee m'}{N};\ \forall m' \in \mathcal{M}_N.$$

Therefore, since $(a+b)^2 \leqslant 2a^2 + 2b^2$ for every $a, b \in \mathbb{R}$, and since $4p(m,\widehat{m}) \leqslant \text{pen}(m) + \text{pen}(\widehat{m})$,

$$\|\widehat{J}_{\widehat{m}} - J_0\|_{\langle M\rangle}^2 \leqslant 3\|\widehat{J}_m - J_0\|_{\langle M\rangle}^2 + 4\text{pen}(m)$$
$$+ 8\left(\left[\sup_{\tau\in\mathcal{B}_{m,\widehat{m}}} |\nu_N(\tau)|\right]^2 - p(m,\widehat{m})\right)_+.$$

By the Bernstein inequality for continuous local martingales (see Revuz and Yor [18], p. 153), for every $\varepsilon > 0$ and $\tau \in \mathbb{L}^2([0,T], d\langle M\rangle_t)$,

$$\mathbb{P}(\nu_N(\tau) \geqslant \varepsilon) \leqslant \exp\left(-\frac{N\varepsilon^2}{2\|\tau\|_{\langle M\rangle}^2}\right).$$

From this inequality, thanks to the same chaining technique as in the proof of Theorem 3.2 (see Step 3),

$$\mathbb{E}\left(\left(\left[\sup_{\tau\in\mathcal{B}_{m,\widehat{m}}} |\nu_N(\tau)|\right]^2 - p(m,\widehat{m})\right)_+\right) \leqslant \frac{\mathfrak{c}_1}{N},$$

where $\mathfrak{c}_1 > 0$ is a constant not depending on N. In conclusion,

$$\mathbb{E}(\|\widehat{J}_{\widehat{m}} - J_0\|_{\langle M\rangle}^2) \leqslant \min_{m\in\mathcal{M}_N}\{3\mathbb{E}(\|\widehat{J}_m - J_0\|_{\langle M\rangle}^2) + 4\text{pen}(m)\} + \frac{8\mathfrak{c}_1}{N}.$$

□

Let us conclude this section with some results on the discrete-time estimator $\widehat{J}_{m,n}$. Proposition 4.15 provides a suitable risk bound on this estimator, but the following technical lemma must be established first.

Lemma 4.1 *Under Assumption 4.3,*

$$\int_0^T \|\Psi_m^{-1}(\varphi_j(s))_j\|_{2,m}^2 d\langle M\rangle_s = \text{trace}(\Psi_m^{-1}) \leqslant \|\mu(\cdot)^{-1}\|_\infty m.$$

Proof Since Ψ_m is a positive definite symmetric matrix,

$$\begin{aligned}
\int_0^T \|\Psi_m^{-1}(\varphi_j(s))_j\|_{2,m}^2 d\langle M\rangle_s &= \sum_{j=1}^m \int_0^T \left(\sum_{\ell=1}^m [\Psi_m^{-1}]_{j,\ell}\varphi_\ell(s)\right)^2 d\langle M\rangle_s \\
&= \sum_{j,\ell,\ell'=1}^m [\Psi_m^{-1}]_{j,\ell}[\Psi_m^{-1}]_{j,\ell'} \int_0^T \varphi_\ell(s)\varphi_{\ell'}(s)d\langle M\rangle_s \\
&= \sum_{j,\ell=1}^m [\Psi_m^{-1}]_{j,\ell} \sum_{\ell'=1}^m [\Psi_m^{-1}]_{j,\ell'}[\Psi_m]_{\ell',\ell} \\
&= \sum_{j,\ell=1}^m [\Psi_m^{-1}]_{j,\ell} I_{j,\ell} = \text{trace}(\Psi_m^{-1})
\end{aligned}$$

and, as in the proof of Lemma 3.1,

$$\begin{aligned}
\|\Psi_m^{-1}\|_{\text{op}} &= \sup_{\theta:\|\Psi_m^{1/2}\theta\|_{2,m}=1} \|\theta\|_{2,m}^2 \\
&= \sup_{J\in S_m:\|J\|_{\langle M\rangle}=1} \|J\|^2 = \sup_{J\in S_m:\|J\|_{\langle M\rangle}=1} \int_0^T \frac{J(s)^2}{\mu(s)} d\langle M\rangle_s \leqslant \|\mu(\cdot)^{-1}\|_\infty.
\end{aligned}$$

Therefore, by Proposition 3.1 (6),

$$\int_0^T \|\Psi_m^{-1}(\varphi_j(s))_j\|_{2,m}^2 d\langle M\rangle_s = \text{trace}(\Psi_m^{-1}) \leqslant \|\Psi_m^{-1}\|_{\text{op}} m \leqslant \|\mu(\cdot)^{-1}\|_\infty m.$$

□

Proposition 4.15 *Under Assumption 4.3, there exists a constant $\mathfrak{c}_{4.2} > 0$, not depending on m, N and n, such that*

$$\mathbb{E}(\|\widehat{J}_{m,n} - J_0\|_{\langle M\rangle}^2) \leqslant 2 \min_{J\in S_m} \|J - J_0\|_{\langle M\rangle}^2 + \mathfrak{c}_{4.2}\left(\frac{m}{N} + \frac{m\mathfrak{R}(m)}{n^2}\right),$$

where

$$\mathfrak{R}(m) = \sup_{t\in[0,T]} \sum_{j=1}^m \varphi_j'(t)^2.$$

Proof First of all, by Proposition 4.14,

$$\begin{aligned}\mathbb{E}(\|\widehat{J}_{m,n} - J_0\|^2_{\langle M\rangle}) &\leqslant 2\mathbb{E}(\|\widehat{J}_m - J_0\|^2_{\langle M\rangle}) + 2\mathbb{E}(\|\widehat{J}_m - \widehat{J}_{m,n}\|^2_{\langle M\rangle})\\ &\leqslant 2\left(\min_{J\in\mathcal{S}_m}\|J - J_0\|^2_{\langle M\rangle} + \frac{2m}{N} + \Delta_{m,n}\right),\end{aligned}$$

where

$$\Delta_{m,n} = \int_0^T \mathbb{E}(\langle \Psi_m^{-1}(\widehat{Z}_m - \widehat{Z}_{m,n}), \varphi(s)\rangle^2_{2,m})d\langle M\rangle_s \quad\text{and}\quad \varphi = (\varphi_1, \dots, \varphi_m).$$

Since $Z^1, \dots, Z^N$ are independent copies of Z, and since Ψ_m^{-1} is a symmetric matrix,

$$\begin{aligned}\Delta_{m,n} &= \int_0^T \mathbb{E}\left[\left(\frac{1}{N}\sum_{i=1}^N\sum_{k=0}^{n-1}\int_{t_k}^{t_{k+1}}\langle \Psi_m^{-1}(\varphi(t) - \varphi(t_k)), \varphi(s)\rangle_{2,m}dZ_t^i\right)^2\right]d\langle M\rangle_s\\ &\leqslant 2\int_0^T\left(\sum_{k=0}^{n-1}\int_{t_k}^{t_{k+1}}\langle\varphi(t) - \varphi(t_k), \Psi_m^{-1}\varphi(s)\rangle_{2,m}J_0(t)d\langle M\rangle_t\right)^2 d\langle M\rangle_s\\ &\quad + 2\int_0^T\mathbb{E}\left[\left(\sum_{k=0}^{n-1}\int_{t_k}^{t_{k+1}}\langle\varphi(t) - \varphi(t_k), \Psi_m^{-1}\varphi(s)\rangle_{2,m}dM_t\right)^2\right]d\langle M\rangle_s\\ &=: 2A_{m,n} + 2B_{m,n}.\end{aligned}$$

Now, let us find suitable bounds on $A_{m,n}$ and $B_{m,n}$:

- **Bound on** $A_{m,n}$. By Cauchy-Schwarz's inequality and Lemma 4.1,

$$\begin{aligned}A_{m,n} &\leqslant \left(\int_0^T \|\Psi_m^{-1}\varphi(s)\|^2_{2,m}d\langle M\rangle_s\right)\left(\sum_{k=0}^{n-1}\int_{t_k}^{t_{k+1}}\|\varphi(t) - \varphi(t_k)\|_{2,m}|J_0(t)|d\langle M\rangle_t\right)^2\\ &\leqslant \operatorname{trace}(\Psi_m^{-1})\|\varphi'\|^2_\infty\left(\sum_{k=0}^{n-1}\int_{t_k}^{t_{k+1}}(t - t_k)|J_0(t)|d\langle M\rangle_t\right)^2\\ &\leqslant \operatorname{trace}(\Psi_m^{-1})\mathfrak{R}(m)\frac{T^2}{n^2}\left(\sum_{k=0}^{n-1}\int_{t_k}^{t_{k+1}}|J_0(t)|d\langle M\rangle_t\right)^2\\ &\leqslant \|\mu(\cdot)^{-1}\|_\infty m\mathfrak{R}(m)\frac{T^2}{n^2}\underbrace{\left(\int_0^T|J_0(t)|d\langle M\rangle_t\right)^2}_{\leqslant\|J_0\|^2_{\langle M\rangle}\langle M\rangle_T} \leqslant \mathfrak{c}_1\frac{m\mathfrak{R}(m)}{n^2}\end{aligned}$$

with

$$\mathfrak{c}_1 = \|\mu(\cdot)^{-1}\|_\infty\|J_0\|^2_{\langle M\rangle}T^2\langle M\rangle_T.$$

- **Bound on** $B_{m,n}$. By the isometry property of Itô's integral, Cauchy-Schwarz's inequality and Lemma 4.1,

$$
\begin{aligned}
B_{m,n} &= \int_0^T \mathbb{E}\left[\left(\int_0^T \left(\sum_{k=0}^{n-1} \langle \varphi(t) - \varphi(t_k), \Psi_m^{-1}\varphi(s)\rangle_{2,m} \mathbf{1}_{[t_k,t_{k+1})}(t)\right) dM_t\right)^2\right] d\langle M\rangle_s \\
&= \int_0^T \int_0^T \left(\sum_{k=0}^{n-1} \langle \varphi(t) - \varphi(t_k), \Psi_m^{-1}\varphi(s)\rangle_{2,m}^2 \mathbf{1}_{[t_k,t_{k+1})}(t)\right) d\langle M\rangle_t d\langle M\rangle_s \\
&\leqslant \left(\int_0^T \|\Psi_m^{-1}\varphi(s)\|_{2,m}^2 d\langle M\rangle_s\right)\left(\sum_{k=0}^{n-1} \int_{t_k}^{t_{k+1}} \|\varphi(t) - \varphi(t_k)\|_{2,m}^2 d\langle M\rangle_t\right) \\
&\leqslant \mathrm{trace}(\Psi_m^{-1})\|\varphi'\|_\infty^2 \sum_{k=0}^{n-1} \int_{t_k}^{t_{k+1}} (t - t_k)^2 d\langle M\rangle_t \\
&\leqslant \mathrm{trace}(\Psi_m^{-1})\mathfrak{R}(m)\frac{T^2}{n^2} \sum_{k=0}^{n-1} \int_{t_k}^{t_{k+1}} d\langle M\rangle_t \leqslant \|\mu(\cdot)^{-1}\|_\infty T^2 \langle M\rangle_T \frac{m\mathfrak{R}(m)}{n^2}.
\end{aligned}
$$

In conclusion, there exists a constant $\mathfrak{c}_2 > 0$, not depending on m, N and n, such that

$$
\mathbb{E}(\|\widehat{J}_{m,n} - J_0\|_{\langle M\rangle}^2) \leqslant 2 \min_{J \in \mathcal{S}_m} \|J - J_0\|_{\langle M\rangle}^2 + \mathfrak{c}_2 \left(\frac{m}{N} + \frac{m\mathfrak{R}(m)}{n^2}\right).
$$

□

For instance, assume that $(\varphi_1, \ldots, \varphi_m)$ is the trigonometric basis. As established in Sect. 4.2.3, $\mathfrak{R}(m)$ is of order m^3. Then, for n of order N^2, the variance term in the risk bound on $\widehat{J}_{m,n}$ provided by Proposition 4.15 is of order m/N, as in the risk bound on the continuous-time estimator of Proposition 4.14.

4.2.4.3 A Brief Reminder on the Fractional Calculus

This section deals with some basics on the (Riemann-Liouville left-sided) fractional integrals and derivatives, which are involved in Sect. 4.2.4.4 for the estimation of b_0 in Eq. (4.20).

Throughout this section, let us consider $\mathtt{l}, \mathtt{r} \in \mathbb{R}$ such that $\mathtt{l} < \mathtt{r}$.

Definition 4.4 The fractional integral of order $\alpha > 0$ of $\varphi \in \mathbb{L}^1([\mathtt{l}, \mathtt{r}])$ is the function $\mathcal{I}_{\mathtt{l}+}^{\alpha}\varphi$ defined by

$$
\mathcal{I}_{\mathtt{l}+}^{\alpha}\varphi(x) = \frac{1}{\Gamma(\alpha)} \int_{\mathtt{l}}^{x} (x - y)^{\alpha-1}\varphi(y)dy;\ \forall x \in [\mathtt{l}, \mathtt{r}].
$$

Definition 4.5 Subject to its existence, the fractional derivative of order $\alpha \in (0, 1)$ of $\varphi : [\mathtt{l}, \mathtt{r}] \to \mathbb{R}$ is the function $\mathcal{D}_{\mathtt{l}+}^{\alpha}\varphi$ defined by

$$
\mathcal{D}_{\mathtt{l}+}^{\alpha}\varphi(x) = \frac{\mathbf{1}_{(\mathtt{l},\mathtt{r})}(x)}{\Gamma(1-\alpha)} \cdot \frac{d}{dx} \int_{\mathtt{l}}^{x} \frac{\varphi(y)}{(x - y)^{\alpha}} dy;\ \forall x \in [\mathtt{l}, \mathtt{r}].
$$

Now, let us consider the space $\mathcal{AC}([1, r])$ of the absolutely continuous functions from $[1, r]$ into $\mathbb{R}$:

$$\mathcal{AC}([1, r]) = \left\{\varphi \in C^0([1, r]) : \exists \varphi' \in \mathbb{L}^1([1, r]), \varphi(\cdot) = \varphi(1) + \int_1^{\cdot} \varphi'(y)dy\right\}.$$

Proposition 4.16 *For every* $\alpha \in (0, 1)$,

$$\mathcal{I}_{1+}^{\alpha}(\mathbb{L}^1([1, r])) = \{\varphi : \mathcal{I}_{1+}^{1-\alpha}(\varphi) \in \mathcal{AC}([1, r])\}.$$

See Samko et al. [19], Theorem 2.3 for a proof.

Proposition 4.17 *Consider* $\varphi \in \mathcal{AC}([1, r])$. *The fractional derivative of order* $\alpha \in (0, 1)$ *of* φ *exists almost everywhere,* $\mathcal{D}_{1+}^{\alpha}\varphi \in \mathbb{L}^{\kappa}([1, r])$ *for every* $\kappa \in [1, 1/\alpha)$, *and*

$$\mathcal{D}_{1+}^{\alpha}\varphi(x) = \frac{1}{\Gamma(1-\alpha)}\left(\frac{\varphi(1)}{(x-1)^{\alpha}} + \int_1^x \frac{\varphi'(y)}{(x-y)^{\alpha}}dy\right); \forall x \in [1, r].$$

See Samko et al. [19], Lemma 2.2 for a proof. Finally, the following theorem provides inversion formulas for the fractional integrals and derivatives.

Theorem 4.6 *Consider* $\alpha \in (0, 1)$. *The fractional operators* $\mathcal{D}_{1+}^{\alpha}$ *and* $\mathcal{I}_{1+}^{\alpha}$ *satisfy the following inversion formulas:*

1. $\mathcal{D}_{1+}^{\alpha}(\mathcal{I}_{1+}^{\alpha}\varphi) = \varphi$ *for every* $\varphi \in \mathbb{L}^1([1, r])$.
2. $\mathcal{I}_{1+}^{\alpha}(\mathcal{D}_{1+}^{\alpha}\varphi) = \varphi$ *for every* $\varphi \in \mathcal{I}_{1+}^{\alpha}(\mathbb{L}^1([1, r]))$.

See Samko et al. [19], Theorem 2.4 for a proof.

4.2.4.4 Application to Some Non-autonomous Fractional Diffusions

In this section, M is the Molchan martingale defined in Sect. 4.2.4.1. For $H = 1/2$, $V : \mathbb{R} \to \mathbb{R}$ is continuously differentiable, V' is bounded and $(b_0, \sigma) : [0, T] \to \mathbb{R} \times \mathbb{R}^*$ is continuous. Then, Eq. (4.20) has a unique solution by Revuz and Yor [18], Chapter IX, Theorem 2.1. For $H \in (1/2, 1)$, $V : \mathbb{R} \to \mathbb{R}$ is twice continuously differentiable, V' and V'' are bounded, $\sigma : [0, T] \to \mathbb{R}^*$ is γ-Hölder continuous with $\gamma \in (1 - H, 1]$, and $b_0 : [0, T] \to \mathbb{R}$ is continuous. Then, Eq. (4.20) has a unique solution whose paths are α-Hölder continuous from $[0, T]$ into $\mathbb{R}$ for every $\alpha \in (1/2, H)$ (see Kubilius et al. [20], Theorem 1.42). In the sequel, the maps V and σ are known and our purpose is to provide a nonparametric estimator of b_0.

First, let us rewrite Eq. (4.20) as a model of type (4.19), thanks to the transformation introduced in Kleptsyna and Le Breton [21] in the parametric estimation framework. Let $Q_0 : [0, T] \to \mathbb{R}$ be the map defined by

$$Q_0(t) = \frac{b_0(t)}{\sigma(t)}; \forall t \in [0, T],$$

and assume that $Q_0 \in \mathcal{Q} = C^1([0, T])$. Consider also the process Z such that, for every $t \in [0, T]$,

$$Z_t = \int_0^t \ell(t, s) dY_s$$

with

$$Y_t = \int_0^t \frac{dX_s}{V(X_s)\sigma(s)} = \int_0^t \left(\frac{b_0(s)}{\sigma(s)} ds + dB_s\right) = \int_0^t Q_0(s) ds + B_t.$$

Then, for any $t \in (0, T]$, Eq. (4.20) leads to

$$\begin{aligned} Z_t &= j(Q_0)(t) + M_t \\ &= \int_0^t J(Q_0)(s) d\langle M \rangle_s + M_t \end{aligned} \tag{4.25}$$

where, for any $Q \in \mathcal{Q}$,

$$j(Q)(t) = \int_0^t \ell(t, s) Q(s) ds \quad \text{and} \quad J(Q)(t) = \frac{1}{2 - 2\mathrm{H}} t^{2\mathrm{H}-1} j(Q)'(t).$$

About the existence of $j(Q)'(t)$, see Kubilius et al. [20], Lemma 5.8.

For $\mathrm{H} = 1/2$, no additional investigations are required because $M = B$ and $J(Q_0) = Q_0$. For $\mathrm{H} \in (1/2, 1)$, in Model (4.25), $\widehat{J}_m$ is a nonparametric estimator of $J(Q_0) \neq Q_0$. So, this section deals with an estimator of Q_0 defined from $\widehat{J}_m$ when $\mathrm{H} \in (1/2, 1)$. Let us consider the function space

$$\mathcal{J} = \left\{\iota : \text{the map } t \in [0, T] \longmapsto \int_0^t s^{1-2\mathrm{H}} \iota(s) ds \text{ belongs to } I_{0+}^{\frac{3}{2}-\mathrm{H}}(\mathbb{L}^1([0, T], dt))\right\}.$$

In order to define our estimator of Q_0, the following technical proposition must be established first.

Proposition 4.18 *The map $J : Q \mapsto J(Q)$ satisfies the two following properties:*

1. $J(\mathcal{Q}) \subset \mathcal{J}$.
2. $\overline{J}(J(Q)) = Q$ *for every $Q \in \mathcal{Q}$, where $\overline{J}$ is the map defined on $\mathcal{J}$ by*

$$\overline{J}(\iota)(t) = \overline{\mathfrak{c}}_{\mathrm{H}} t^{\mathrm{H}-\frac{1}{2}} \int_0^t (t - s)^{\mathrm{H}-\frac{3}{2}} s^{1-2\mathrm{H}} \iota(s) ds; \forall (\iota, t) \in \mathcal{J} \times [0, T]$$

$$\textit{with } \overline{\mathfrak{c}}_{\mathrm{H}} = \frac{2 - 2\mathrm{H}}{\mathfrak{c}_{\mathrm{H}} \Gamma(3/2 - \mathrm{H}) \Gamma(\mathrm{H} - 1/2)}.$$

Proof Consider $Q \in \mathcal{Q}$, and the function $Q_{\mathrm{H}} : (0, T] \to \mathbb{R}$ defined by

$$Q_{\mathrm{H}}(t) = t^{\frac{1}{2}-\mathrm{H}} Q(t); \forall t \in (0, T].$$

The function $\mathcal{I}_{0+}^{3/2-\mathrm{H}}(Q_{\mathrm{H}})$ is well-defined on $(0, T]$ and, for every $t \in (0, T]$,

$$\begin{aligned}
\mathcal{I}_{0+}^{\frac{3}{2}-\mathrm{H}}(Q_{\mathrm{H}})(t) &= \frac{1}{\Gamma(3/2-\mathrm{H})} \int_0^t (t-s)^{\frac{1}{2}-\mathrm{H}} Q_{\mathrm{H}}(s) ds \\
&= \frac{1}{\mathfrak{c}_{\mathrm{H}}\Gamma(3/2-\mathrm{H})} \int_0^t \ell(t,s) Q(s) ds = \frac{1}{\mathfrak{c}_{\mathrm{H}}\Gamma(3/2-\mathrm{H})} j(Q)(t).
\end{aligned}$$

Since $j(Q)'(t)$ exists for any $t \in (0, T]$ by Kubilius et al. [20], Lemma 5.8 as mentioned above, the derivative of $\mathcal{I}_{0+}^{3/2-\mathrm{H}}(Q_{\mathrm{H}})$ at time t also. Moreover, since Q is continuous on $[0, T]$,

$$\begin{aligned}
|\mathcal{I}_{0+}^{\frac{3}{2}-\mathrm{H}}(Q_{\mathrm{H}})(t)| &\leqslant \frac{1}{\Gamma(3/2-\mathrm{H})} \int_0^t s^{\frac{1}{2}-\mathrm{H}} (t-s)^{\frac{1}{2}-\mathrm{H}} |Q(s)| ds \\
&\leqslant \frac{\|Q\|_{\infty}}{2\Gamma(3/2-\mathrm{H})} \left(\int_0^t s^{1-2\mathrm{H}} ds + \int_0^t (t-s)^{1-2\mathrm{H}} ds \right) \\
&= \frac{\|Q\|_{\infty}}{(2-2\mathrm{H})\Gamma(3/2-\mathrm{H})} t^{2-2\mathrm{H}} \xrightarrow[t \to 0]{} 0.
\end{aligned}$$

By the definition of the map J, for every $s \in (0, T]$,

$$\begin{aligned}
J(Q)(s) &= \frac{1}{2-2\mathrm{H}} s^{2\mathrm{H}-1} j(Q)'(s) \\
&= \frac{\mathfrak{c}_{\mathrm{H}}\Gamma(3/2-\mathrm{H})}{2-2\mathrm{H}} s^{2\mathrm{H}-1} \frac{\partial}{\partial s} \mathcal{I}_{0+}^{\frac{3}{2}-\mathrm{H}}(Q_{\mathrm{H}})(s)
\end{aligned}$$

and then, for every $t \in [0, T]$,

$$\begin{aligned}
\int_0^t s^{1-2\mathrm{H}} J(Q)(s) ds &= \frac{\mathfrak{c}_{\mathrm{H}}\Gamma(3/2-\mathrm{H})}{2-2\mathrm{H}} \lim_{\varepsilon \to 0} \int_{\varepsilon}^t \left(\frac{\partial}{\partial s} \mathcal{I}_{0+}^{\frac{3}{2}-\mathrm{H}}(Q_{\mathrm{H}})(s) \right) ds \\
&= \frac{\mathfrak{c}_{\mathrm{H}}\Gamma(3/2-\mathrm{H})}{2-2\mathrm{H}} \lim_{\varepsilon \to 0} (\mathcal{I}_{0+}^{\frac{3}{2}-\mathrm{H}}(Q_{\mathrm{H}})(t) - \mathcal{I}_{0+}^{\frac{3}{2}-\mathrm{H}}(Q_{\mathrm{H}})(\varepsilon)) \\
&= \frac{\mathfrak{c}_{\mathrm{H}}\Gamma(3/2-\mathrm{H})}{2-2\mathrm{H}} \mathcal{I}_{0+}^{\frac{3}{2}-\mathrm{H}}(Q_{\mathrm{H}})(t). \qquad (4.26)
\end{aligned}$$

So, $J(Q) \in \mathcal{J}$ by Equality (4.26), and then $J(\mathcal{Q}) \subset \mathcal{J}$. By applying the fractional derivative of order $3/2 - \mathrm{H}$ on each side of Equality (4.26), and thanks to its representation for absolutely continuous functions on $[0, T]$ (see Proposition 4.17), for every $t \in (0, T]$,

$$Q_{\mathrm{H}}(t) = \frac{2-2\mathrm{H}}{\mathfrak{c}_{\mathrm{H}}\Gamma(3/2-\mathrm{H})} \mathcal{D}_{0+}^{\frac{3}{2}-\mathrm{H}} \left(\int_0^{\cdot} s^{1-2\mathrm{H}} J(Q)(s) ds \right)(t)$$

$$= \overline{\mathfrak{c}}_{\mathrm{H}} \int_0^t (t-s)^{\mathrm{H}-\frac{3}{2}} s^{1-2\mathrm{H}} J(Q)(s)ds.$$

Therefore, for every $t \in [0, T]$,

$$Q(t) = \overline{\mathfrak{c}}_{\mathrm{H}} t^{\mathrm{H}-\frac{1}{2}} \int_0^t (t-s)^{\mathrm{H}-\frac{3}{2}} s^{1-2\mathrm{H}} J(Q)(s)ds = \overline{J}(J(Q))(t).$$

□

For $\varphi_1, \dots, \varphi_m \in \mathcal{J}$, Proposition 4.18 legitimates to consider the following estimator of Q_0:

$$\widehat{Q}_m(t) = \overline{J}(\widehat{J}_m)(t); \, t \in [0, T].$$

The following proposition provides risk bounds on $\widehat{Q}_m$ ($m \in \{1, \dots, N\}$), and on the adaptive estimator $\widehat{Q}_{\widehat{m}}$.

Proposition 4.19 *Assume that $J(Q_0) \in \mathbb{L}^2([0, T], d\langle M\rangle_t)$. If the φ_j's belong to $\mathcal{J}$, then there exists a constant $\mathfrak{c}_{4.19,1} > 0$, not depending on N, such that*

$$\mathbb{E}(\|\widehat{Q}_m - Q_0\|^2) \leqslant \mathfrak{c}_{4.19,1} \left(\min_{\iota \in \mathcal{S}_m} \|\iota - J(Q_0)\|_{\langle M\rangle}^2 + \frac{m}{N} \right); \forall m \in \{1, \dots, N\}.$$

Moreover, under Assumption 4.4, there exists a constant $\mathfrak{c}_{4.19,2} > 0$, not depending on N, such that

$$\mathbb{E}(\|\widehat{Q}_{\widehat{m}} - Q_0\|^2) \leqslant \mathfrak{c}_{4.19,2} \left(\min_{m \in \mathcal{M}_N} \left\{ \min_{\iota \in \mathcal{S}_m} \|\iota - J(Q_0)\|_{\langle M\rangle}^2 + \frac{m}{N} \right\} + \frac{1}{N} \right).$$

Proof Consider $\iota \in \mathcal{J} \cap \mathbb{L}^2([0, T], d\langle M\rangle_t)$. By Cauchy-Schwarz's inequality,

$$\begin{aligned} \|\overline{J}(\iota)\|^2 &= \overline{\mathfrak{c}}_{\mathrm{H}}^2 \int_0^T t^{2\mathrm{H}-1} \left(\int_0^t (t-s)^{\mathrm{H}-\frac{3}{2}} s^{1-2\mathrm{H}} \iota(s)ds \right)^2 dt \\ &\leqslant \overline{\mathfrak{c}}_{\mathrm{H}}^2 \int_0^T t^{2\mathrm{H}-1} \theta(t) \int_0^t (t-s)^{\mathrm{H}-\frac{3}{2}} s^{1-2\mathrm{H}} \iota(s)^2 ds dt \end{aligned}$$

with

$$\theta(t) = \int_0^t (t-s)^{\mathrm{H}-\frac{3}{2}} s^{1-2\mathrm{H}} ds; \, \forall t \in (0, T].$$

Note that $\overline{\theta} : t \mapsto t^{2\mathrm{H}-1}\theta(t)$ is bounded on $(0, T]$. Indeed, for every $t \in (0, T]$,

$$|\overline{\theta}(t)| \leqslant t^{2\mathrm{H}-1} \left[\left(\frac{t}{2}\right)^{\mathrm{H}-\frac{3}{2}} \int_0^{t/2} s^{1-2\mathrm{H}} ds + \left(\frac{t}{2}\right)^{1-2\mathrm{H}} \int_{t/2}^t (t-s)^{\mathrm{H}-\frac{3}{2}} ds \right]$$

$$= t^{2\mathrm{H}-1}\left[\frac{1}{2-2\mathrm{H}}\left(\frac{t}{2}\right)^{\mathrm{H}-\frac{3}{2}}\left(\frac{t}{2}\right)^{2-2\mathrm{H}} + \frac{1}{H-1/2}\left(\frac{t}{2}\right)^{1-2\mathrm{H}}\left(\frac{t}{2}\right)^{\mathrm{H}-\frac{1}{2}}\right]$$
$$= \frac{1}{2^{1/2-\mathrm{H}}}\left(\frac{1}{2-2\mathrm{H}} + \frac{1}{\mathrm{H}-1/2}\right)t^{\mathrm{H}-\frac{1}{2}} \xrightarrow[t\to 0^+]{} 0.$$

So, by the Fubini-Tonelli theorem,

$$\|\overline{J}(\iota)\|^2 \leqslant \overline{\mathfrak{c}}_{\mathrm{H}}^2\|\overline{\theta}\|_\infty \int_0^T\int_0^T (t-s)^{\mathrm{H}-\frac{3}{2}}s^{1-2\mathrm{H}}\iota(s)^2\mathbf{1}_{(s,T]}(t)dsdt$$
$$= \overline{\mathfrak{c}}_{\mathrm{H}}^2\|\overline{\theta}\|_\infty \int_0^T s^{1-2\mathrm{H}}\iota(s)^2\int_s^T (t-s)^{\mathrm{H}-\frac{3}{2}}dtds$$
$$\leqslant \frac{\overline{\mathfrak{c}}_{\mathrm{H}}^2}{\mathrm{H}-1/2}T^{\mathrm{H}-\frac{1}{2}}\|\overline{\theta}\|_\infty \int_0^T s^{1-2\mathrm{H}}\iota(s)^2 ds.$$

Then, $\overline{J}(\iota) \in \mathbb{L}^2([0,T],dt)$ and

$$\|\overline{J}(\iota)\|^2 \leqslant \mathfrak{c}_1\|\iota\|_{\langle M\rangle}^2 \quad \text{with} \quad \mathfrak{c}_1 = \frac{\overline{\mathfrak{c}}_{\mathrm{H}}^2}{(\mathrm{H}-1/2)(2-2\mathrm{H})}T^{\mathrm{H}-\frac{1}{2}}\|\overline{\theta}\|_\infty. \tag{4.27}$$

Therefore, by the definition of the estimator $\widehat{Q}_m$, Proposition 4.18 and Inequality (4.27),

$$\|\widehat{Q}_m - Q_0\|^2 = \|\overline{J}(\widehat{J}_m) - \overline{J}(J(Q_0))\|^2$$
$$= \|\overline{J}(\widehat{J}_m - J(Q_0))\|^2 \leqslant \mathfrak{c}_1\|\widehat{J}_m - J(Q_0)\|_{\langle M\rangle}^2.$$

The conclusion follows from Proposition 4.14 and Theorem 4.5.

□

Proposition 4.19 says that the MISE (Mean Integrated Squared Error) of $\widehat{Q}_m$ ($m \in \{1,\dots,N\}$) (resp. $\widehat{Q}_{\widehat{m}}$) has at most the same bound as the MISE of $\widehat{J}_m$ (resp. $\widehat{J}_{\widehat{m}}$). In order to easily check that the φ_j's belong to $\mathcal{J}$ in Proposition 4.19, let us provide a simple subset of $\mathcal{J}$.

Proposition 4.20 *The function space*

$$\mathbb{J} = \left\{\iota \in C^1([0,T]) : \lim_{t\to 0^+} t^{-2\mathrm{H}}\iota(t) \text{ and } \lim_{t\to 0^+} t^{1-2\mathrm{H}}\iota'(t) \text{ exist and are unique}\right\} \quad \text{is a subset of } \mathcal{J}.$$

Proof First, by Proposition 4.16,

$$\mathcal{J} = \left\{\iota : I_{0+}^{\mathrm{H}-\frac{1}{2}}\left(\int_0^{\cdot} s^{1-2\mathrm{H}}\iota(s)ds\right) \in \mathcal{AC}([0,T])\right\},$$

and then

$$\mathcal{J}_{\rm Lip} = \left\{\iota \in C^0([0,T]) : \mathcal{I}_{0+}^{\rm H-\frac{1}{2}}\left(\int_0^{\cdot} s^{1-2\rm H}\iota(s)ds\right) \text{ is Lipschitz continuous}\right\}$$

is a subset of $\mathcal{J}$. For every $\iota \in C^0([0,T])$ and $t \in [0,T]$, since

$$\int_{[0,t]^2} |(t-u)^{\rm H-\frac{3}{2}} s^{1-2\rm H}\iota(s)\mathbf{1}_{(0,u)}(s)|dsdu$$
$$\leqslant \frac{\|\iota\|_\infty}{2-2\rm H}\int_0^t (t-u)^{\rm H-\frac{3}{2}}u^{2-2\rm H}du \leqslant \frac{\|\iota\|_\infty T^{2-2\rm H}}{2-2\rm H}\underbrace{\int_0^t (t-u)^{\rm H-\frac{3}{2}}du}_{=(\rm H-1/2)^{-1}t^{\rm H-1/2}} < \infty,$$

by the Fubini-Lebesgue theorem,

$$\begin{aligned}\mathcal{I}_{0+}^{\rm H-\frac{1}{2}}\left(\int_0^{\cdot} s^{1-2\rm H}\iota(s)ds\right)(t) &= \frac{1}{\Gamma(\rm H-1/2)}\int_0^t (t-u)^{\rm H-\frac{3}{2}}\int_0^u s^{1-2\rm H}\iota(s)dsdu\\ &= \frac{1}{\Gamma(\rm H-1/2)}\int_0^t s^{1-2\rm H}\iota(s)\int_s^t (t-u)^{\rm H-\frac{3}{2}}duds\\ &= \frac{1}{(\rm H-1/2)\Gamma(\rm H-1/2)}\int_0^t (t-s)^{\rm H-\frac{1}{2}}s^{1-2\rm H}\iota(s)ds.\end{aligned}$$

Then,

$$\mathcal{J}_{\rm Lip} = \left\{\iota \in C^0([0,T]) : t \mapsto \int_0^t (t-s)^{\rm H-\frac{1}{2}}s^{1-2\rm H}\iota(s)ds \text{ is Lipschitz continuous}\right\}.$$

Now, let us show that $\mathbb{J} \subset \mathcal{J}_{\rm Lip}$. For any $\iota \in \mathbb{J}$ and every $t \in [0,T]$,

$$\begin{aligned}\varphi(t) &= \int_0^t (t-s)^{\rm H-\frac{1}{2}}s^{1-2\rm H}\iota(s)ds\\ &= \lim_{\varepsilon\to 0}\frac{(t-\varepsilon)^{\rm H+1/2}}{\rm H+1/2}\varepsilon^{1-2\rm H}\iota(\varepsilon)\\ &\quad + \int_0^t \frac{(t-s)^{\rm H+1/2}}{\rm H+1/2}((1-2\rm H)s^{-2\rm H}\iota(s) + s^{1-2\rm H}\iota'(s))ds\\ &= \int_0^t \theta(s)(t-s)^{\rm H+\frac{1}{2}}ds,\end{aligned}$$

where

$$\theta(s) = \frac{1}{\rm H+1/2}((1-2\rm H)s^{-2\rm H}\iota(s) + s^{1-2\rm H}\iota'(s))\ ;\ \forall s \in (0,T].$$

Since $\iota \in \mathbb{J}$, the map $\theta : t \mapsto \theta(t)$ is bounded on $(0, T]$, and for every $t, \bar{t} \in [0, T]$ such that $\bar{t} > t$,

$$\begin{aligned}
|\varphi(\bar{t}) - \varphi(t)| &\leqslant \int_t^{\bar{t}} |\theta(s)|(\bar{t} - s)^{\mathrm{H}+\frac{1}{2}} ds \\
&\quad + \int_0^t |\theta(s)| \cdot |(\bar{t} - s)^{\mathrm{H}+\frac{1}{2}} - (t - s)^{\mathrm{H}+\frac{1}{2}}| ds \\
&\leqslant \|\theta\|_\infty T^{\mathrm{H}+\frac{1}{2}} |\bar{t} - t| + \left(\mathrm{H} + \frac{1}{2}\right) |\bar{t} - t| \int_0^t \sup_{u \in [t-s, \bar{t}-s]} u^{\mathrm{H}-\frac{1}{2}} ds \\
&\leqslant \left(\mathrm{H} + \frac{3}{2}\right) \|\theta\|_\infty T^{\mathrm{H}+\frac{1}{2}} |\bar{t} - t|.
\end{aligned}$$

So, the function $t \mapsto \varphi(t)$ is Lipschitz continuous from $[0, T]$ into $\mathbb{R}$, and then $\iota \in \mathcal{J}_{\mathrm{Lip}}$. In conclusion, $\mathbb{J} \subset \mathcal{J}_{\mathrm{Lip}} \subset \mathcal{J}$.

□

Consider $\psi_1, \dots, \psi_m \in C^1([0, T])$ such that $(\psi_1, \dots, \psi_m)$ is an orthonormal family of $\mathbb{L}^2([0, T], dt)$. In particular, $\psi_1, \dots, \psi_m$ are linearly independent. Let us assume that the basis $(\varphi_1, \dots, \varphi_m)$ of $\mathcal{S}_m$, orthonormal in $\mathbb{L}^2([0, T], dt)$, is obtained via the Gram-Schmidt process from $(\overline{\varphi}_1, \dots, \overline{\varphi}_m)$, where $\overline{\varphi}_j(t) = t^{2\mathrm{H}} \psi_j(t)$ for every $j \in \{1, \dots, m\}$ and $t \in [0, T]$. Then, the φ_j's belong to $\mathbb{J} \subset \mathcal{J}$. First, let us establish a suitable bound on the bias term of our estimator of Q. Consider the functions $\upsilon : [0, T] \to \mathbb{R}$ and $\overline{\mu} : (0, T] \to \mathbb{R}$, defined by

$$\upsilon(t) := t^{2\mathrm{H}} \mu(t)^{\frac{1}{2}} = (2 - 2\mathrm{H})^{\frac{1}{2}} t^{\mathrm{H}+\frac{1}{2}} \quad \text{and} \quad \overline{\mu}(t) := \upsilon(t)^{-1} \mu(t)^{\frac{1}{2}} = t^{-2\mathrm{H}}.$$

For every $\iota \in \mathcal{S}_m$, there exist $\alpha_1, \dots, \alpha_m \in \mathbb{R}$ such that

$$\iota = \sum_{j=1}^m \alpha_j \overline{\varphi}_j = \upsilon \mu^{-\frac{1}{2}} \bar{\iota}$$

with

$$\bar{\iota} = \sum_{j=1}^m \alpha_j \psi_j \in \mathbb{S}_m \quad \text{and} \quad \mathbb{S}_m = \mathrm{span}\{\psi_1, \dots, \psi_m\}.$$

Thus, by assuming that $\overline{\mu} J(Q_0) \in \mathbb{L}^2([0, T], dt)$,

$$\begin{aligned}
\min_{\iota \in \mathcal{S}_m} \|\iota - J(Q_0)\|_{\langle M \rangle}^2 &= \min_{\bar{\iota} \in \mathbb{S}_m} \|\upsilon(\bar{\iota} - \upsilon^{-1} \mu^{\frac{1}{2}} J(Q_0))\|^2 \\
&\leqslant (2 - 2\mathrm{H}) T^{2\mathrm{H}+1} \min_{\bar{\iota} \in \mathbb{S}_m} \|\bar{\iota} - \overline{\mu} J(Q_0)\|^2 \\
&= (2 - 2\mathrm{H}) T^{2\mathrm{H}+1} \|\Pi_{\mathbb{S}_m}(\overline{\mu} J(Q_0)) - \overline{\mu} J(Q_0)\|^2,
\end{aligned}$$

where $\Pi_{\mathbb{S}_m}$ is the orthogonal projection from $\mathbb{L}^2([0,T],dt)$ onto $\mathbb{S}_m$. If $(\psi_1,\dots,\psi_m)$ is the $[0,T]$-supported trigonometric basis, and if $\overline{\mu}J(Q_0)\in\mathbb{W}_2^{\beta}([0,T])$ with $\beta\in\mathbb{N}^*$, then

$$\min_{\iota\in S_m}\|\iota - J(Q_0)\|_{\langle M\rangle}^2 \leqslant \mathfrak{c}_{\beta,T}(2-2\mathrm{H})T^{2\mathrm{H}+1}m^{-2\beta}.$$

Now, consider $\kappa>2\mathrm{H}$, and let us show that, for instance, $\overline{\mu}J(Q_0)\in\mathbb{L}^2([0,T],dt)$ when $Q_0(t)=t^{\kappa}$ for any $t\in[0,T]$. By the change of variable formula,

$$\begin{aligned}
j(Q_0)(t) &= \int_0^t \ell(t,s)Q_0(s)ds\\
&= \mathfrak{c}_{\mathrm{H}}\int_0^t s^{\kappa+\frac{1}{2}-\mathrm{H}}(t-s)^{\frac{1}{2}-\mathrm{H}}ds\\
&= \mathfrak{c}_{\mathrm{H}}\int_{\varepsilon}^t (t-s)^{\kappa+\frac{1}{2}-\mathrm{H}}s^{\frac{1}{2}-\mathrm{H}}ds\\
&\quad + \int_0^{\varepsilon}\underbrace{\mathfrak{c}_{\mathrm{H}}(t-s)^{\kappa+\frac{1}{2}-\mathrm{H}}s^{\frac{1}{2}-\mathrm{H}}}_{=:\varphi(s,t)}ds \quad\text{with}\quad \varepsilon>0.
\end{aligned}$$

Then, by Lebesgue's theorem,

$$\begin{aligned}
j(Q_0)'(t) &= \varphi(t,t)\cdot 1 - \varphi(\varepsilon,t)\cdot 0 + \int_0^t \partial_t\varphi(s,t)ds\\
&= \overline{\mathfrak{c}}_{\mathrm{H}}\int_0^t (t-s)^{\kappa-\frac{1}{2}-\mathrm{H}}s^{\frac{1}{2}-\mathrm{H}}ds \quad\text{with}\quad \overline{\mathfrak{c}}_{\mathrm{H}} = \mathfrak{c}_{\mathrm{H}}\left(\kappa+\frac{1}{2}-\mathrm{H}\right).
\end{aligned}$$

Therefore, for every $t\in(0,T]$,

$$\begin{aligned}
|(\overline{\mu}J(Q_0))(t)| &= \frac{\overline{\mathfrak{c}}_{\mathrm{H}}}{2-2\mathrm{H}}t^{-1}\int_0^t (t-s)^{\kappa-\frac{1}{2}-\mathrm{H}}s^{\frac{1}{2}-\mathrm{H}}ds\\
&\leqslant \frac{\overline{\mathfrak{c}}_{\mathrm{H}}}{2-2\mathrm{H}}t^{\kappa-\frac{3}{2}-\mathrm{H}}\int_0^t s^{\frac{1}{2}-\mathrm{H}}ds \leqslant \frac{\overline{\mathfrak{c}}_{\mathrm{H}}}{2-2\mathrm{H}}\left(\frac{3}{2}-\mathrm{H}\right)^{-1}T^{\kappa-2\mathrm{H}}.
\end{aligned}$$

Let us conclude this section with the estimation of the drift function for two models of type (4.20) in finance.

Example 4.1 Let us consider a financial market model in which the prices process $S=(S_t)_{t\in\mathbb{R}_+}$ of the risky asset satisfies

$$S_t = S_0 + \int_0^t S_u(b_0(u)du + \sigma dW_u);\; t\in\mathbb{R}_+ \tag{4.28}$$

where S_0 is a $(0, \infty)$-valued random variable, $W = (W_t)_{t\in\mathbb{R}_+}$ is a Brownian motion, $\sigma > 0$ and $b_0 \in C^0(\mathbb{R}_+)$. This is a non-autonomous extension of the Black-Scholes model. Note that, in practice, several independent copies of the prices process S cannot be observed on $[0, T]$. So, in order to define a suitable estimator of b_0 on $[0, T]$, let us assume that S is observed on $[0, NT]$ and, for any $i \in \{1, \dots, N\}$, consider $T_i = (i-1)T$ and the process $S^i = (S_t^i)_{t\in[0,T]}$ defined by

$$S_t^i = S_{T_i+t};\ \forall t \in [0, T].$$

By Eq. (4.28), for every $t \in [0, T]$,

$$\begin{aligned} S_t^i &= S_{T_i} + \int_{T_i}^{T_i+t} S_u(b_0(u)du + \sigma dW_u) \\ &= S_0^i + \int_0^t S_u^i(b_0(T_i+u)du + \sigma dW_u^i) \quad \text{with} \quad W^i = W_{T_i+\cdot} - W_{T_i}. \end{aligned}$$

Moreover, let $Z^i = (Z_t^i)_{t\in[0,T]}$ be the process such that, for every $t \in [0, T]$,

$$Z_t^i = \frac{1}{\sigma}\int_0^t \frac{dS_u^i}{S_u^i} = \frac{1}{\sigma}\int_0^t b_0(T_i+u)d\langle W^i\rangle_u + W_t^i. \tag{4.29}$$

Since $W^1, \dots, W^N$ are N independent Brownian motions, by assuming that the volatility constant σ is known and that b_0 is T-periodic, a suitable nonparametric estimator of b_0 on $[0, T]$ is given by

$$\widehat{b}_m(t) = \sigma\widehat{J}_m(t);\ t \in [0, T],$$

where $m \in \{1, \dots, N\}$ and

$$\widehat{J}_m \in \arg\min_{J\in\mathcal{S}_m}\left\{\frac{1}{N}\sum_{i=1}^N\left(\int_0^T J(s)^2ds - 2\int_0^T J(s)dZ_s^i\right)\right\}.$$

Since

$$\|\widehat{b}_m - b_0\|^2 = \sigma^2\|\widehat{J}_m - J_0\|^2 \quad \text{with} \quad J_0 = \frac{b_0}{\sigma}.$$

Proposition 4.14 provides a risk bound on $\widehat{b}_m$, and Theorem 4.5 provides a risk bound on $\widehat{b}_{\widehat{m}}$ with $\widehat{m}$ selected in $\mathcal{M}_N \subset \{1, \dots, N\}$ via (4.24). Finally, to assume that b_0 is T-periodic means that Model (4.28) is appropriate for assets with a prices process having *similar* trends on each interval $[T_i, T_{i+1}]$, typically each day ($T =$ 24h). Obviously, since constant functions are T-periodic, $\widehat{b}_m$ is an estimator of the drift constant in the usual Black-Scholes model.

Now, assume that $S_0 = 10$, $\sigma = 0.2$ and that

$$b_0(t) = \sin(2\pi t) + \cos(2\pi t);\ \forall t \in \mathbb{R}_+.$$

For $N = 100$ days, let us simulate one path of the prices process S along the dissection $\{100\ell/n;\ \ell = 0, \dots, n\}$ of $[0, 100]$ with $n = N^2 = 10000$. The simulated path of S and those of $Z^1, \dots, Z^N$ are plotted on Fig. 4.5.

The adaptive estimator $\widehat{b}_{\widehat{m}}$ is computed from $Z^1, \dots, Z^N$ in the $[0, 1]$-supported $\widehat{m}$-dimensional trigonometric basis with $\widehat{m}$ selected in $\{1, \dots, 12\}$. This experiment is repeated 100 times, and 10 adaptive estimations of b_0 are plotted on Fig. 4.6. On average, the MISE of $\widehat{b}_{\widehat{m}}$ is very small: 0.002.

Example 4.2 Let us consider a financial market model in which the prices process $S = (S_t)_{t\in[0,T]}$ and the volatility process $\sigma = (\sigma_t)_{t\in[0,T]}$ of the risky asset satisfy

$$\begin{cases} dS_t = S_t(b(t)dt + \sigma_t dW_t) \\ d\sigma_t = \sigma_t(\rho_0(t)dt + \upsilon dB_t) \end{cases} \tag{4.30}$$

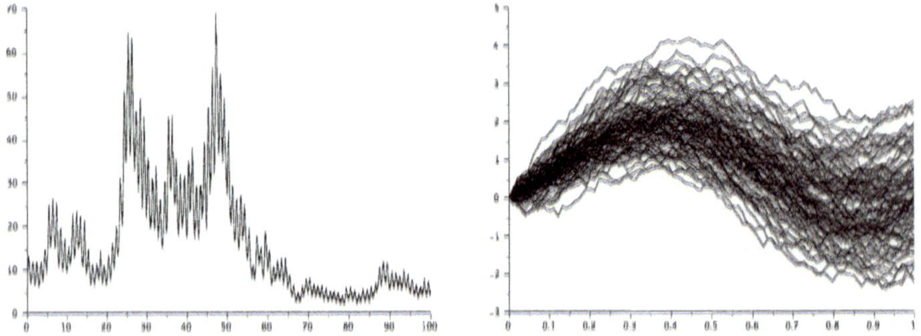

Fig. 4.5 Plots of the paths of $Z^1, \dots, Z^N$ (right) computed from one path of the non-autonomous Black-Scholes model (left)

Fig. 4.6 Plots of 10 adaptive estimations (black dashed lines) of b_0 (red line)

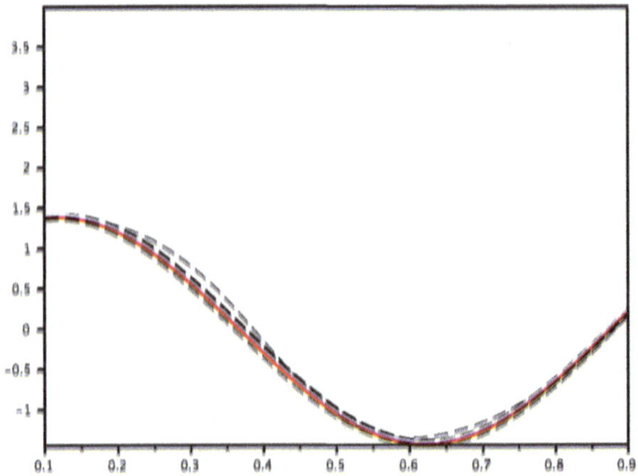

where S_0 and σ_0 are $(0,\infty)$-valued random variables, $W = (W_t)_{t\in\mathbb{R}_+}$ (resp. $B = (B_t)_{t\in\mathbb{R}_+}$) is a Brownian motion (resp. a fractional Brownian motion of Hurst parameter $\mathrm{H} \in [1/2, 1)$), $\upsilon > 0$ and $b, \rho_0 \in C^0(\mathbb{R}_+)$. Note that if $\mathrm{H} > 1/2$, then Model (4.30) takes into account the persistence-in-volatility phenomenon as in Comte et al. [22]. Here again, in practice, it is not possible to get several independent copies of the volatility process σ on $[0, T]$. So, in order to define a suitable estimator of ρ_0 on $[0, T]$, let us assume that (S, σ) is observed on $[0, N(T + \Delta)]$ with $\Delta \in \mathbb{R}_+$, and for any $i \in \{1, \dots, N\}$, consider $T_i(\Delta) = (i - 1)(T + \Delta)$ and the process $\sigma^i = (\sigma^i_t)_{t\in[0,T]}$ defined by

$$\sigma^i_t = \sigma_{T_i(\Delta)+t}; \ \forall t \in [0, T].$$

By Eq. (4.30), for every $t \in [0, T]$,

$$\begin{aligned}
\sigma^i_t &= \sigma_{T_i(\Delta)} + \int_{T_i(\Delta)}^{T_i(\Delta)+t} \sigma_s(\rho_0(s)ds + \upsilon dB_s)\\
&= \sigma^i_0 + \int_0^t \sigma^i_s(\rho_0(T_i(\Delta) + s)ds + \upsilon dB^i_s) \quad \text{with} \quad B^i = B_{T_i(\Delta)+\cdot} - B_{T_i(\Delta)}.
\end{aligned}$$

Moreover, let $Z^i = (Z^i_t)_{t\in[0,T]}$ be the process such that, for every $t \in [0, T]$,

$$\begin{aligned}
Z^i_t &= \frac{\mathfrak{c}_{\mathrm{H}}}{\upsilon}\int_0^t s^{\frac{1}{2}-\mathrm{H}}(t-s)^{\frac{1}{2}-\mathrm{H}}\frac{d\sigma^i_s}{\sigma^i_s}\\
&= \frac{1}{\upsilon}\int_0^t \ell(t, s)\rho_0(T_i(\Delta) + s)ds + M^i_t,
\end{aligned}$$

where

$$M^i_t = \int_0^t \ell(t, s)dB^i_s; \ \forall t \in [0, T].$$

In the sequel, ρ_0 is $(T + \Delta)$-periodic, and then

$$Z^i_t = \frac{1}{\upsilon}\int_0^t J(\rho_0)(s)d\langle M^i\rangle_s + M^i_t; \ \forall t \in [0, T].$$

Since B has stationary increments, $M^1, \dots, M^N$ have the same distribution, but these Molchan martingales are not independent when $\mathrm{H} > 1/2$. However, for any $i, k \in \{1, \dots, N\}$ such that $i < k$, and any $s, t \in [0, T]$ such that $s < t$,

$$\begin{aligned}
\mathbb{E}(B^i_s B^k_t) &= \mathbb{E}(B_s(B_{t+T_{i,k}(\Delta)} - B_{T_{i,k}(\Delta)})) \quad \text{with} \quad T_{i,k}(\Delta) = T_k(\Delta) - T_i(\Delta)\\
&= \frac{1}{2}(s^{2\mathrm{H}} + (t + T_{i,k}(\Delta))^{2\mathrm{H}} - (t + T_{i,k}(\Delta) - s)^{2\mathrm{H}}\\
&\quad - (s^{2\mathrm{H}} + T_{i,k}(\Delta)^{2\mathrm{H}} - (T_{i,k}(\Delta) - s)^{2\mathrm{H}}))\\
&= \frac{1}{2}((T_{i,k}(\Delta) + t)^{2\mathrm{H}} + (T_{i,k}(\Delta) - s)^{2\mathrm{H}} - (T_{i,k}(\Delta) + t - s)^{2\mathrm{H}} - T_{i,k}(\Delta)^{2\mathrm{H}})
\end{aligned}$$

$$= \frac{1}{2} T_{i,k}(\Delta)^{2\mathrm{H}}((1 + t/T_{i,k}(\Delta))^{2\mathrm{H}} + (1 - s/T_{i,k}(\Delta))^{2\mathrm{H}} - (1 + (t-s)/T_{i,k}(\Delta))^{2\mathrm{H}} - 1)$$

$$= \frac{1}{2} \mathrm{H}(2\mathrm{H}-1) T_{i,k}(\Delta)^{2\mathrm{H}}((t/T_{i,k}(\Delta))^2 + (s/T_{i,k}(\Delta))^2 - ((t-s)/T_{i,k}(\Delta))^2 + \mathrm{o}((1/T_{i,k}(\Delta))^2)) \quad \text{when} \quad \Delta \to \infty.$$

Thus,

$$\mathbb{E}(B_s^i B_t^k) \underset{\Delta \to \infty}{\sim} \mathrm{H}(2\mathrm{H}-1) \cdot st \cdot (k-i)^{2\mathrm{H}-2}(T+\Delta)^{2\mathrm{H}-2},$$

and since $(T + \Delta)^{2\mathrm{H}-2} \to 0$ when $\Delta \to \infty$, the larger Δ, the more B^i and B^k (and then M^i and M^k) *become* independent. So, for Δ large enough, if the constant υ is known, a *satisfactory* nonparametric estimator of ρ_0 is given by

$$\widehat{\rho}_m(t) = \upsilon \widehat{Q}_m(t);\ t \in [0, T],$$

where $m \in \{1, \dots, N\}$,

$$\widehat{Q}_m(t) = \begin{cases} \widehat{J}_m(t) & \text{if } \mathrm{H} = 1/2 \\ \overline{\mathfrak{c}}_{\mathrm{H}} t^{\mathrm{H}-\frac{1}{2}} \displaystyle\int_0^t (t-s)^{\mathrm{H}-\frac{3}{2}} s^{1-2\mathrm{H}} \widehat{J}_m(s) ds & \text{if } \mathrm{H} > 1/2 \end{cases} ;\ t \in [0, T]$$

and

$$\widehat{J}_m = \arg\min_{J \in \mathcal{S}_m} \left\{ \frac{1}{N} \sum_{i=1}^{N} \left((2 - 2\mathrm{H}) \int_0^T J(s)^2 s^{1-2\mathrm{H}} ds - 2 \int_0^T J(s) dZ_s^i \right) \right\}.$$

Assume that the φ_j's belong to $\mathcal{J}$. Since

$$\|\widehat{\rho}_m - \rho_0\|^2 = \upsilon^2 \|\widehat{Q}_m - Q_0\|^2 \quad \text{with} \quad Q_0 = \frac{\rho_0}{\upsilon},$$

if $M^1, \dots, M^N$ were independent, then Proposition 4.19 would provide risk bounds on $\widehat{\rho}_m$ and on the adaptive estimator $\widehat{\rho}_{\widehat{m}}$ with $\widehat{m}$ selected in $\mathcal{M}_N$ via (4.24). Of course $M^1, \dots, M^N$ are not independent when $\mathrm{H} > 1/2$, but the risk bounds of Proposition 4.19 remain consistent for Δ large enough as explained above. Finally, the $(T + \Delta)$-periodicity condition on ρ_0 makes sense in the following special case: when ρ_0 is T-periodic and $\Delta = \delta T$ with $\delta \in \mathbb{N}^*$ large enough. As in Example 4.1, to assume that ρ_0 is T-periodic means that Model (4.30) is appropriate for assets with a volatility process having *similar* trends on each interval $[(i-1)T, iT]$, typically each day ($T = 24\mathrm{h}$). When $T = 24\mathrm{h}$, to assume δ large enough means to skip enough days between two days during which the volatility process is observed in order to estimate ρ_0 with our method.

References

1. Applebaum, D. (2009). *Lévy Processes and Stochastic Calculus*. Cambridge University Press, Cambridge.
2. Privault, N. (2022). *Introduction to Stochastic Finance with Market Examples*. Chapman & Hall/CRC, New York.
3. Chen, Z.Q., Hu, E., Xie, L. and Zhang, X. (2017). Heat Kernels for Nonsymmetric Diffusion Operators with Jumps. *Journal of Differential Equations* **263**(10), 6576–6634.
4. Amorino, C. and Gloter, A. (2021). Invariant Density Adaptive Estimation for Ergodic Jump Diffusion Processes over Anisotropic Classes. *Journal of Statistical Planning and Inference* **213**, 106–129.
5. Abramowitz, M. and Stegun, I.A. (1964). *Handbook of Mathematical Functions with Formulas, Graphs, and Mathematical Tables*. Dover Publications, USA.
6. Comte, F. and Genon-Catalot, V. (2020). Regression Function Estimation as a Partly Inverse Problem. *The Annals of the Institute of Statistical Mathematics* **72**(4), 1023–1054.
7. Belomestny, D., Comte, F. and Genon-Catalot, V. (2019). Sobolev-Hermite versus Sobolev Nonparametric Density Estimation on $\mathbb{R}$. *The Annals of the Institute of Statistical Mathematics* **71**(1), 29–62.
8. Comte, F. (2001). Adaptive Estimation of the Spectrum of a Stationary Gaussian Sequence. *Bernoulli* **7**(2), 267–298.
9. Comte, F. and Genon-Catalot, V. (2020). Nonparametric Drift Estimation for I.I.D. Paths of Stochastic Differential Equations. *The Annals of Statistics* **48**(6), 3336–3365.
10. Friz, P. and Victoir, N. (2010). *Multidimensional Stochastic Processes as Rough Paths: Theory and Applications*. Cambridge University Press, Cambridge.
11. Nualart, D. (2006). *The Malliavin Calculus and Related Topics*. Springer, Berlin-Heidelberg.
12. Biagini, F., Hu, Y., Oksendal, B. and Zhang, T. (2008). *Stochastic Calculus for Fractional Brownian Motion and Applications*. Springer, London.
13. Hu, Y., Nualart, D. and Zhou, H. (2019). Drift Parameter Estimation for Nonlinear Stochastic Differential Equations Driven by Fractional Brownian Motion. *Stochastics* **91**(8), 1067–1091.
14. Comte, F. and Marie, N. (2021). Nonparametric Estimation for I.I.D. Paths of Fractional SDE. *Statistical Inference for Stochastic Processes* **24**(3), 669–705.
15. Decreusefond, L. (2022). *Selected Topics in Malliavin Calculus*. Bocconi & Springer Series.
16. Li, X-M., Panloup, F. and Sieber, J. (2023). On the (Non-)Stationary Density of Fractional-Driven Stochastic Differential Equations. *The Annals of Probability* **51**(6), 2056–2085.
17. DeVore, R.A. and Lorentz, G.G. (1993). *Constructive Approximation*. Springer, Berlin-Heidelberg.
18. Revuz, D. and Yor, M. (1999). *Continuous Martingales and Brownian Motion*. Springer, Berlin-Heidelberg.
19. Samko, S.G., Kilbas, A.A. and Marichev, O.I. (1993). *Fractional Integrals and Derivatives. Theory and Applications*. CRC Press, Boca Raton.
20. Kubilius, K., Mishura, Y. and Ralchenko, K. (2017). *Parameter Estimation in Fractional Diffusion Models*. Bocconi & Springer Series.
21. Kleptsyna, M.L. and Le Breton, A. (2001). Some Explicit Statistical Results about Elementary Fractional Type Models. *Nonlinear Analysis* **47**, 4783–4794.
22. Comte, F., Coutin, L. and Renault, E. (2012). Affine Fractional Stochastic Volatility Models. *The Annals of Finance* **8**, 337–378.

Chapter 5
The Nadaraya-Watson Estimator of the Drift Function

This chapter deals with risk bounds on the following 2-bandwidths (truncated) Nadaraya-Watson estimator of b_0:

$$\widehat{b}_{h,\mathrm{h}}(x) = \frac{\widehat{bf}_h(x)}{\widehat{f}_{\mathrm{h}}(x)} \mathbf{1}_{\widehat{f}_{\mathrm{h}}(x) > \frac{m}{2}} \; ; \; x \in [\mathtt{l}, \mathtt{r}],$$

where $h, \mathrm{h} > 0$, $m \in (0, 1]$ and $\mathtt{l}, \mathtt{r} \in \mathbb{R}$ satisfy $\mathtt{l} < \mathtt{r}$. First of all, Sect. 5.1 deals with a Nikol'skii type condition on the density function f. In Sect. 5.2, risk bounds on $\widehat{f}_{\mathrm{h}}$, on $\widehat{bf}_h$, and then on $\widehat{b}_{h,\mathrm{h}}$ are established for fixed h and h. Finally, Sect. 5.3 deals with an extension of the PCO method to select h and h from data, and then with risk bounds on the associated adaptive estimators of f, $b_0 f$ and b_0.

5.1 A Nikol'skii Type Condition on the Time-Average Diffusion Density

As mentioned in the beginning of Chap. 3, if σ satisfies the non-degeneracy condition (1.3), then the probability distribution of X_t ($t \in (0, T]$) has a density f_t with respect to Lebesgue measure such that

$$f_t(x) \leqslant \mathfrak{c}_{0.5,T} t^{-\frac{1}{2}} \exp\left(-\mathfrak{m}_{0.5,T} \frac{(x - x_0)^2}{t}\right) \; ; \; \forall x \in \mathbb{R}.$$

For any $t_0 \in (0, T)$, this legitimates considering the density function f, already defined in Chap. 1 by

$$f(x) = \frac{1}{T - t_0} \int_{t_0}^{T} f_s(x) ds \; ; \; x \in \mathbb{R}.$$

N. Marie, *From Nonparametric Regression to Statistical Inference for Non-Ergodic Diffusion Processes*, Frontiers in Probability and the Statistical Sciences,
https://doi.org/10.1007/978-3-031-95638-6_5

In order to establish a suitable bound on the bias term of $\widehat{f}_h$ in Sect. 5.2, the f_t's must be at least one time continuously differentiable, and their derivatives need to satisfy a bound of type (3.1). Precisely, in the sequel, the f_t's fulfill the following assumption.

Assumption 5.1 There exists $\beta \in \mathbb{N}^*$ such that, for any $t \in (0, T]$, f_t is β times continuously differentiable. Moreover, for any $x \in \mathbb{R}$,

$$|f_t^{(\ell)}(x)| \leqslant \frac{\mathfrak{c}_{5.1}(\ell)}{t^{q(\ell)}} \exp\left(-\mathfrak{m}_{5.1}(\ell)\frac{(x-x_0)^2}{t}\right) \ ; \forall \ell \in \{1, \dots, \beta\},$$

where all the constants are positive, depend on T, but not on t and x.

Notation. For every $p \in \overline{\mathbb{N}}\backslash\{0\}$, $C_{\mathrm{b}}^p(\mathbb{R})$ is the space of all the p times continuously differentiable functions $\varphi : \mathbb{R} \to \mathbb{R}$ such that

$$\|\varphi^{(k)}\|_\infty < \infty; \forall k \in \{0, \dots, p\}.$$

In the two following situations, the f_t's fulfill Assumption 5.1:

- Assume that the functions b_0 and σ belong to $C_{\mathrm{b}}^\infty(\mathbb{R})$. Since σ satisfies (1.3), by Kusuoka and Stroock [1], Corollary 3.25, the f_t's fulfill Assumption 5.1.
- Assume that b_0 is Lipschitz continuous (but not bounded), and that $\sigma \in C_{\mathrm{b}}^1(\mathbb{R})$. Assume also that σ' is Hölder continuous. Then, by Menozzi et al. [2], Theorem 1.2, the f_t's fulfill Assumption 5.1 with $\beta = 1$.

Under Assumption 5.1, for any $\kappa \geqslant 1$ and any continuous function $\varphi : \mathbb{R} \to \mathbb{R}$ having polynomial growth, $t \mapsto \mathbb{E}(|\varphi(X_t)|^\kappa)$ is bounded on $[0, T]$. Indeed, for any $t \in [0, T]$,

$$\begin{aligned}\mathbb{E}(|\varphi(X_t)|^\kappa) &\leqslant \mathfrak{c}_1(1 + \mathbb{E}(|X_t|^{q\kappa})) = \mathfrak{c}_1 \int_{-\infty}^{\infty} (1 + |x|^{q\kappa}) f_t(x)dx \\ &\leqslant \mathfrak{c}_1\mathfrak{c}_{0.5,T} \int_{-\infty}^{\infty} (1 + |t^{\frac{1}{2}}x + x_0|^{q\kappa})e^{-\mathfrak{m}_{0.5,T}x^2}dx \leqslant \mathfrak{c}_2(1 \vee T^{\frac{q\kappa}{2}}),\end{aligned}$$

where

$$\mathfrak{c}_2 = \mathfrak{c}_1\mathfrak{c}_{0.5,T} \int_{-\infty}^{\infty} (1 + (|x| + |x_0|)^{q\kappa})e^{-\mathfrak{m}_{0.5,T}x^2}dx,$$

and the constants $\mathfrak{c}_1, q > 0$ only depend on φ. Moreover,

$$\begin{aligned}\|\varphi^\kappa\|_f^2 &= \int_{-\infty}^{\infty} |\varphi(x)|^\kappa f(x)dx \\ &= \frac{1}{T - t_0} \int_{t_0}^{T} \mathbb{E}(|\varphi(X_s)|^\kappa)ds \leqslant \mathfrak{c}_2(1 \vee T^{\frac{q\kappa}{2}}).\end{aligned}$$

Then, $\varphi^\kappa \in \mathbb{L}^2(\mathbb{R}, f(x)dx)$ and $\|\varphi^\kappa\|_f$ is bounded by a constant which doesn't depend on t_0. For instance, this applies to b_0 and σ with $q = 1$.

Now, let us show that f satisfies a Nikol'skii type condition.

Proposition 5.1 *For every $x \in \mathbb{R}$, $f(x) > 0$. Moreover, under Assumption 5.1, there exists a constant $\mathfrak{c}_{5.1} > 0$, depending on T but not on t_0, such that for every $\ell \in \{0, \dots, \beta - 1\}$ and $\theta \in \mathbb{R}$,*

$$\int_{-\infty}^{\infty} (f^{(\ell)}(\theta + x) - f^{(\ell)}(x))^2 dx \leqslant \frac{\mathfrak{c}_{5.1}}{t_0^{2q(\ell+1)}}(\theta^2 + |\theta|^3).$$

Proof First of all, since $f_t(x) > 0$ for every $(t, x) \in (0, T] \times \mathbb{R}$,

$$f(x) = \frac{1}{T - t_0} \int_{t_0}^{T} f_t(x)dt > 0; \forall x \in \mathbb{R}.$$

Consider $\ell \in \{1, \dots, \beta\}$ and $\theta \in \mathbb{R}_+$. Thanks to the bound on $(t, x) \mapsto f_t^{(\ell)}(x)$ given in Assumption 5.1,

$$\begin{aligned}
&\|f^{(\ell-1)}(\theta + \cdot) - f^{(\ell-1)}\|^2 \\
&\leqslant \frac{1}{T - t_0} \int_{t_0}^{T} \int_{-\infty}^{\infty} (f_t^{(\ell-1)}(\theta + x + x_0) - f_t^{(\ell-1)}(x + x_0))^2 dxdt \\
&\leqslant \frac{\theta^2}{T - t_0} \int_{t_0}^{T} \int_{-\infty}^{\infty} \sup_{z \in [x, x+\theta]} f_t^{(\ell)}(z + x_0)^2 dxdt \\
&\leqslant \mathfrak{c}_{5.1}(\ell)^2 \frac{\theta^2}{T - t_0} \int_{t_0}^{T} \frac{1}{t^{2q(\ell)}} \int_{-\infty}^{\infty} \sup_{z \in [x, x+\theta]} \exp\left(-2\mathfrak{m}_{5.1}(\ell)\frac{z^2}{t}\right) dxdt \\
&= \mathfrak{c}_{5.1}(\ell)^2 \frac{\theta^2}{T - t_0} \int_{t_0}^{T} \frac{1}{t^{2q(\ell)}} \left(\int_{-\infty}^{-\theta} \exp\left(-2\mathfrak{m}_{5.1}(\ell)\frac{(x + \theta)^2}{t}\right) dx + \theta \right. \\
&\qquad \left. + \int_{0}^{\infty} \exp\left(-2\mathfrak{m}_{5.1}(\ell)\frac{x^2}{t}\right) dx \right) dt \\
&\leqslant \frac{1}{t_0^{2q(\ell)}} \left(\mathfrak{c}_1\theta^2 + \theta^3 \max_{k \in \{1, \dots, \beta\}} \mathfrak{c}_{5.1}(k)^2\right)
\end{aligned}$$

with

$$\mathfrak{c}_1 = 2 \max_{k \in \{1, \dots, \beta\}} \left\{\mathfrak{c}_{5.1}(k)^2 \int_{0}^{\infty} \exp\left(-2\mathfrak{m}_{5.1}(k)\frac{x^2}{T}\right) dx\right\}.$$

In the same way,

$$\begin{aligned}
&\|f^{(\ell-1)}(-\theta + \cdot) - f^{(\ell-1)}\|^2 \\
&\leqslant \frac{\theta^2}{T - t_0} \int_{t_0}^{T} \int_{-\infty}^{\infty} \sup_{z \in [x-\theta, x]} f_t^{(\ell)}(z + x_0)^2 dxdt
\end{aligned}$$

$$\leqslant \mathfrak{c}_{5.1}(\ell)^2 \frac{\theta^2}{T-t_0} \int_{t_0}^{T} \frac{1}{t^{2q(\ell)}} \left(\int_{-\infty}^{0} \exp\left(-2\mathfrak{m}_{5.1}(\ell)\frac{x^2}{t}\right) dx + \theta \right.$$
$$\left. + \int_{\theta}^{\infty} \exp\left(-2\mathfrak{m}_{5.1}(\ell)\frac{(x-\theta)^2}{t}\right) dx \right) dt$$
$$\leqslant \frac{1}{t_0^{2q(\ell)}} \left(\mathfrak{c}_1\theta^2 + \theta^3 \max_{k\in\{1,\dots,\beta\}} \mathfrak{c}_{5.1}(k)^2 \right).$$

This concludes the proof. □

Assumption 5.1 and Proposition 5.1 are crucial in the sequel, but t_0 needs to be chosen carefully to get suitable risk bounds on $\widehat{f}_h$ and $\widehat{bf}_h$. Since the probability distribution of X_t at time $t = 0$ is a Dirac measure while it has a smooth density with respect to Lebesgue's measure for every $t \in (0, T]$, the behavior of the bounds on $(t, x) \mapsto f_t(x)$ and its derivatives is singular at point $(0, x_0)$. Consequently, all the bounds in Sect. 5.2 depend on t_0 through a multiplicative constant of order $1/\min\{t_0^{\alpha}, T - t_0\}$ with $\alpha > 0$, and then one should take $t_0 \in [1, T - 1]$ when $T > 1$.

5.2 Nonadaptive Risk Bounds

This section deals with risk bounds on $\widehat{f}_h$, on $\widehat{bf}_h$, and then on the Nadaraya-Watson estimator $\widehat{b}_h$. In the sequel, the kernel K fulfills the following usual assumptions.

Assumption 5.2 The kernel K is symmetric, continuous and belongs to $\mathbb{L}^2(\mathbb{R}, dx)$.

Assumption 5.3 There exists $\upsilon \in \mathbb{N}^*$ such that

$$\int_{-\infty}^{\infty} |z^{\upsilon+1} K(z)| dz < \infty$$

and

$$\int_{-\infty}^{\infty} z^{\ell} K(z) dz = 0; \forall \ell \in \{1, \dots, \upsilon\}.$$

About the construction of kernels fulfilling both Assumptions 5.2 and 5.3, the reader may refer to Comte [3], Proposition 2.10. The following proposition provides a risk bound on $\widehat{f}_h$.

Proposition 5.2 *Under Assumption 5.2,*

$$\mathbb{E}(\|\widehat{f}_h - f\|^2) \leqslant \|f - f_h\|^2 + \frac{\|K\|^2}{Nh} \quad \text{with} \quad f_h = K_h * f.$$

Moreover, under Assumptions 5.1 and 5.3 with $\upsilon = \beta$,

$$\|f - f_h\|^2 \leqslant \mathfrak{c}_{5.2}(t_0)h^{2\beta},$$

where

$$\mathfrak{c}_{5.1}(t_0) = \frac{\mathfrak{c}_{5.2}}{t_0^{2q(\beta)}[\mathbf{1}_{\beta=1} + [(\beta-2)!]^2\mathbf{1}_{\beta\geqslant 2}]}\left(\int_{-\infty}^{\infty} |z|^{\beta}(1+|z|^{\frac{1}{2}})|K(z)|dz\right)^2.$$

Proof First of all, let us recall that

$$\mathbb{E}(\|\widehat{f}_h - f\|^2) = \int_{-\infty}^{\infty} \mathfrak{b}(x)^2 dx + \int_{-\infty}^{\infty} \mathfrak{v}(x)dx$$

where, for any $x \in \mathbb{R}$, $\mathfrak{b}(x)$ is the bias of $\widehat{f}_h(x)$ and $\mathfrak{v}(x)$ is its variance. On the one hand, let us find a suitable bound on the integrated variance of $\widehat{f}_h$. Since $X^1, \dots, X^N$ are i.i.d. copies of X, and thanks to Jensen's inequality,

$$\begin{aligned}
\mathfrak{v}(x) &= \text{var}\left[\frac{1}{N(T-t_0)}\sum_{i=1}^{N}\int_{t_0}^{T} K_h(X_s^i - x)ds\right]\\
&= \frac{1}{N(T-t_0)^2}\text{var}\left(\int_{t_0}^{T} K_h(X_s - x)ds\right) \leqslant \frac{1}{N}\mathbb{E}\left[\left(\int_{t_0}^{T} K_h(X_s - x)\frac{ds}{T-t_0}\right)^2\right]\\
&\leqslant \frac{1}{N(T-t_0)}\int_{t_0}^{T}\mathbb{E}(K_h(X_s - x)^2)ds = \frac{1}{N}\int_{-\infty}^{\infty} K_h(z-x)^2 f(z)dz.
\end{aligned}$$

Thus, since K is symmetric,

$$\begin{aligned}
\int_{-\infty}^{\infty}\mathfrak{v}(x)dx &\leqslant \frac{1}{N}\int_{-\infty}^{\infty} f(z)\int_{-\infty}^{\infty} K_h(z-x)^2 dxdz\\
&= \frac{1}{Nh}\left(\int_{-\infty}^{\infty} f(z)dz\right)\left(\int_{-\infty}^{\infty} K(x)^2 dx\right) = \frac{\|K\|^2}{Nh}.
\end{aligned}$$

On the other hand, let us find a suitable bound on the integrated squared-bias of $\widehat{f}_h$. Since $X^1, \dots, X^N$ are i.i.d. copies of X,

$$\begin{aligned}
\mathfrak{b}(x) &= \frac{1}{T-t_0}\int_{t_0}^{T}\mathbb{E}(K_h(X_s - x))ds - f(x)\\
&= \frac{1}{h}\int_{-\infty}^{\infty} K\left(\frac{z-x}{h}\right) f(z)dz - f(x)\\
&= \int_{-\infty}^{\infty} K(z)(f(hz+x) - f(x))dz.
\end{aligned}$$

First, assume that $\beta = 1$. By Assumption 5.3, the generalized Minkowski inequality (see Comte [3], Theorem B.1) and Proposition 5.1,

$$\int_{-\infty}^{\infty} \mathfrak{b}(x)^2 dx \leqslant \int_{-\infty}^{\infty} \left(\int_{-\infty}^{\infty} K(z)(f(hz+x) - f(x))dz \right)^2 dx$$
$$\leqslant \left[\int_{-\infty}^{\infty} |K(z)| \left(\int_{-\infty}^{\infty} (f(hz+x) - f(x))^2 dx \right)^{\frac{1}{2}} dz \right]^2 \leqslant \mathfrak{c}_1(t_0) h^2$$

with

$$\mathfrak{c}_1(t_0) = \frac{\mathfrak{c}_{5.1}}{t_0^{2q(1)}} \left(\int_{-\infty}^{\infty} |z|(1 + |z|^{\frac{1}{2}})|K(z)|dz \right)^2.$$

Now, assume that $\beta \geqslant 2$. By the Taylor formula with integral remainder, for every $z \in \mathbb{R}$,

$$f(hz+x) - f(x) = \mathbf{1}_{\beta \geqslant 3} \sum_{\ell=1}^{\beta-2} \frac{(hz)^{\ell}}{\ell!} f^{(\ell)}(x)$$
$$+ \frac{(hz)^{\beta-1}}{(\beta-2)!} \int_0^1 (1-\tau)^{\beta-2} f^{(\beta-1)}(\tau hz + x) d\tau.$$

Then, by Assumption 5.3, the generalized Minkowski inequality and by Proposition 5.1,

$$\int_{-\infty}^{\infty} \mathfrak{b}(x)^2 dx = \int_{-\infty}^{\infty} \left(\int_{-\infty}^{\infty} K(z)(f(hz+x) - f(x))dz \right)^2 dx$$
$$= \left(\frac{h^{\beta-1}}{(\beta-2)!} \right)^2 \int_{-\infty}^{\infty} \left(\int_{-\infty}^{\infty} z^{\beta-1} K(z) \right.$$
$$\left. \times \int_0^1 (1-\tau)^{\beta-2} (f^{(\beta-1)}(\tau hz + x) - f^{(\beta-1)}(x)) d\tau dz \right)^2 dx$$
$$\leqslant \left(\frac{h^{\beta-1}}{(\beta-2)!} \right)^2 [\int_{-\infty}^{\infty} |z|^{\beta-1} |K(z)|$$
$$\times \int_0^1 (1-\tau)^{\beta-2} \left(\int_{-\infty}^{\infty} (f^{(\beta-1)}(\tau hz + x) - f^{(\beta-1)}(x))^2 dx \right)^{\frac{1}{2}} d\tau dz]^2$$

$$\leqslant \frac{\mathfrak{c}_{5.1} h^{2\beta}}{[t_0^{q(\beta)}(\beta-2)!]^2} \left(\int_{-\infty}^{\infty} |z|^{\beta-1} |K(z)| \right.$$
$$\left. \times \int_0^1 (1-\tau)^{\beta-2} (\tau|z| + (\tau|z|)^{\frac{3}{2}}) d\tau dz \right)^2 \leqslant \mathfrak{c}_2(t_0) h^{2\beta}$$

with

$$\mathfrak{c}_2(t_0) = \frac{\mathfrak{c}_{5.1}}{[t_0^{q(\beta)}(\beta-2)!]^2}\left(\int_{-\infty}^{\infty}|z|^{\beta}(1+|z|^{\frac{1}{2}})|K(z)|dz\right)^2.$$

This concludes the proof. □

Note that thanks to Proposition 5.2, the bias-variance tradeoff is reached by (the risk bound on) $\widehat{f}_h$ when h is of order $N^{-1/(2\beta+1)}$, leading to a rate of order $N^{-2\beta/(2\beta+1)}$. Moreover, to take $t_0 \geqslant 1$ when $T > 1$ gives

$$\mathbb{E}(\|\widehat{f}_h - f\|^2) \leqslant \mathfrak{c}_{5.2}h^{2\beta} + \frac{\|K\|^2}{Nh}$$
$$\text{with} \quad \mathfrak{c}_{5.2} = \frac{\mathfrak{c}_{5.1}}{\mathbf{1}_{\beta=1} + [(\beta-2)!]^2\mathbf{1}_{\beta\geqslant 2}}\left(\int_{-\infty}^{\infty}|z|^{\beta}(1+|z|^{\frac{1}{2}})|K(z)|dz\right)^2.$$

The following proposition provides a risk bound on $\widehat{bf}_h$.

Proposition 5.3 *Under Assumption 5.2,*

$$\mathbb{E}(\|\widehat{bf}_h - b_0 f\|^2) \leqslant \|(bf)_h - b_0 f\|^2 + \frac{\mathfrak{c}_{5.3}(t_0)}{Nh}$$

with

$$(bf)_h = K_h * (b_0 f) \quad \textit{and} \quad \mathfrak{c}_{5.3}(t_0) = 2\|K\|^2\left(\|b_0\|_f^2 + \frac{1}{T-t_0}\|\sigma\|_f^2\right).$$

Proof First of all, let us recall that

$$\mathbb{E}(\|\widehat{bf}_h - b_0 f\|^2) = \int_{-\infty}^{\infty}\mathfrak{b}(x)^2dx + \int_{-\infty}^{\infty}\mathfrak{v}(x)dx$$

where, for any $x \in \mathbb{R}$, $\mathfrak{b}(x)$ is the bias of $\widehat{bf}_h(x)$ and $\mathfrak{v}(x)$ is its variance. On the one hand, let us find a suitable bound on the integrated variance of $\widehat{bf}_h$. Since $X^1, \dots, X^N$ are i.i.d. copies of X,

$$\begin{aligned}\mathfrak{v}(x) &= \frac{1}{N(T-t_0)^2}\text{var}\left(\int_{t_0}^{T}K_h(X_s - x)dX_s\right)\\ &\leqslant \frac{2}{N}\mathbb{E}\left[\left(\int_{t_0}^{T}K_h(X_s - x)b_0(X_s)\frac{ds}{T-t_0}\right)^2\right.\\ &\qquad\qquad \left. + \frac{1}{(T-t_0)^2}\left(\int_{t_0}^{T}K_h(X_s - x)\sigma(X_s)dW_s\right)^2\right].\end{aligned}$$

Then, by Jensen's inequality and the isometry property of Itô's integral,

$$\begin{aligned}\mathfrak{v}(x) &\leqslant \frac{2}{N(T-t_0)}\int_{t_0}^{T}\mathbb{E}(K_h(X_s-x)^2 b_0(X_s)^2)ds\\ &\qquad + \frac{2}{N(T-t_0)^2}\int_{t_0}^{T}\mathbb{E}(K_h(X_s-x)^2\sigma(X_s)^2)ds\\ &= \frac{2}{N}\int_{-\infty}^{\infty}K_h(z-x)^2 b_0(z)^2 f(z)dz + \frac{2}{N(T-t_0)}\int_{-\infty}^{\infty}K_h(z-x)^2\sigma(z)^2 f(z)dz.\end{aligned}$$

Since K is symmetric and $K \in \mathbb{L}^2(\mathbb{R}, dx)$ by Assumption 5.2, and since $b_0, \sigma \in \mathbb{L}^2(\mathbb{R}, f(x)dx)$,

$$\begin{aligned}\int_{-\infty}^{\infty}\mathfrak{v}(x)dx &\leqslant \frac{2}{N}\int_{\mathbb{R}^2}K_h(z-x)^2 b_0(z)^2 f(z)dzdx\\ &\qquad + \frac{2}{N(T-t_0)}\int_{\mathbb{R}^2}K_h(z-x)^2\sigma(z)^2 f(z)dzdx\\ &\leqslant \frac{2\|K\|^2}{Nh}\left(\int_{-\infty}^{\infty}b_0(z)^2 f(z)dz + \frac{1}{T-t_0}\int_{-\infty}^{\infty}\sigma(z)^2 f(z)dz\right).\end{aligned}$$

On the other hand, let us find a suitable bound on the integrated squared-bias of $\widehat{bf}_h$. Again, since $X^1, \dots, X^N$ are i.i.d. copies of X, and since Itô's integral restricted to $\mathbb{H}^2$ is a martingale-valued map,

$$\begin{aligned}\mathfrak{b}(x) &= \mathbb{E}\left[\frac{1}{N(T-t_0)}\sum_{i=1}^{N}\int_{t_0}^{T}K_h(X_s^i-x)dX_s^i\right] - b_0(x)f(x)\\ &= \frac{1}{T-t_0}\left(\mathbb{E}\left(\int_{t_0}^{T}K_h(X_s-x)b_0(X_s)ds\right)\right.\\ &\qquad \left. + \mathbb{E}\left(\int_{t_0}^{T}K_h(X_s-x)\sigma(X_s)dW_s\right)\right) - b_0(x)f(x)\\ &= \int_{-\infty}^{\infty}K_h(z-x)b_0(z)f(z)dz - b_0(x)f(x) = ((bf)_h - b_0 f)(x).\end{aligned}$$

Therefore, since f is bounded and b_0 belongs to $\mathbb{L}^2(\mathbb{R}, f(x)dx)$,

$$\int_{-\infty}^{\infty}\mathfrak{b}(x)^2 dx = \|(bf)_h - b_0 f\|^2.$$

This concludes the proof.

□

Consider $\gamma \in \mathbb{N}^*$. Assume that $b_0 f$ is γ times continuously differentiable, and that there exists $\varphi \in \mathbb{L}^1(\mathbb{R}, |x|^{\gamma-1}K(x)dx)$ such that, for every $\theta \in \mathbb{R}$ and $h \in (0, 1]$,

$$\int_{-\infty}^{\infty}((b_0 f)^{(\gamma-1)}(h\theta + x) - (b_0 f)^{(\gamma-1)}(x))^2 dx \leqslant \varphi(\theta)h^2. \tag{5.1}$$

If in addition K fulfills Assumption 5.3 with $\upsilon = \gamma$, then $\|(bf)_h - b_0 f\|^2$ is of order $h^{2\gamma}$, and by Proposition 5.3, the bias-variance tradeoff is reached by $\widehat{bf}_h$ when h is of order $N^{-1/(2\gamma+1)}$, leading to a rate of order $N^{-2\gamma/(2\gamma+1)}$. Moreover, to take $t_0 \leqslant T - 1$ when $T > 1$ gives

$$\mathbb{E}(\|\widehat{bf}_h - b_0 f\|^2) \leqslant \|(bf)_h - b_0 f\|^2 + \frac{\mathfrak{c}_{5.3}}{Nh}$$
$$\text{with} \quad \mathfrak{c}_{5.3} = 2\|K\|^2(\|b_0\|_f^2 + \|\sigma\|_f^2).$$

Finally, Propositions 5.2 and 5.3 allow to provide a risk bound on a truncated version of the Nadaraya-Watson estimator $\widehat{b}_h$.

Notation. For every $\varphi \in \mathbb{L}^2(\mathbb{R}, f(x)dx)$, $\|\varphi\|_{f,\mathtt{l},\mathtt{r}} = \|\varphi \mathbf{1}_{[\mathtt{l},\mathtt{r}]}\|_f$.

Proposition 5.4 *Consider the 2-bandwidths Nadaraya-Watson estimator*

$$\widehat{b}_{h,\ \mathtt{h}} = \frac{\widehat{bf}_h}{\widehat{f}_{\mathtt{h}}} \mathbf{1}_{\widehat{f}_{\mathtt{h}}(\cdot) > \frac{m}{2}} \quad \textit{with} \quad h,\quad \mathtt{h} > 0,$$

and assume that $f(x) > m$ *for every* $x \in [\ \mathtt{l},\ \mathtt{r}]$ *(*$m \in (0, 1]$*). Under Assumptions 5.1, 5.2 and 5.3 with* $\upsilon = \beta$*,*

$$\mathbb{E}(\|\widehat{b}_{h,\ \mathtt{h}} - b_0\|_{f,\ \mathtt{l},\ \mathtt{r}}^2) \leqslant \frac{\mathfrak{c}_{5.4}}{m^2}\left(\|(bf)_h - b_0 f\|^2 + \frac{\mathfrak{c}_{5.3}(t_0)}{Nh}\right.$$
$$\left. + \|b_0\|_f^2\left(\mathfrak{c}_{5.2}(t_0)\ \mathtt{h}^{2\beta} + \frac{\|K\|^2}{N\ \mathtt{h}}\right)\right),$$

where $\mathfrak{c}_{5.4} = 16(\|f\|_\infty \vee \|b_0^2 f\|_\infty)$.

Proof First of all,

$$\widehat{b}_{h,\mathtt{h}} - b_0 = \left(\frac{\widehat{bf}_h - b_0 f}{\widehat{f}_{\mathtt{h}}} + \left(\frac{1}{\widehat{f}_{\mathtt{h}}} - \frac{1}{f}\right) b_0 f\right) \mathbf{1}_{\widehat{f}_{\mathtt{h}}(\cdot) > \frac{m}{2}} - b_0 \mathbf{1}_{\widehat{f}_{\mathtt{h}}(\cdot) \leqslant \frac{m}{2}}.$$

Then,

$$\|\widehat{b}_{h,\mathtt{h}} - b_0\|_{f,\mathtt{l},\mathtt{r}}^2 = \|b_0 \mathbf{1}_{\widehat{f}_{\mathtt{h}}(\cdot) \leqslant \frac{m}{2}}\|_{f,\mathtt{l},\mathtt{r}}^2$$
$$+ \left\|\left(\frac{\widehat{bf}_h - b_0 f}{\widehat{f}_{\mathtt{h}}} + \left(\frac{1}{\widehat{f}_{\mathtt{h}}} - \frac{1}{f}\right) b_0 f\right) \mathbf{1}_{\widehat{f}_{\mathtt{h}}(\cdot) > \frac{m}{2}}\right\|_{f,\mathtt{l},\mathtt{r}}^2.$$

Moreover, for any $x \in [\mathtt{l}, \mathtt{r}]$, since $f(x) > m$, for every $\omega \in \{\widehat{f}_{\mathtt{h}}(\cdot) \leqslant m/2\}$,

$$|f(x) - \widehat{f}_{\mathrm{h}}(x,\omega)| \geqslant f(x) - \widehat{f}_{\mathrm{h}}(x,\omega) > m - \frac{m}{2} = \frac{m}{2}.$$

Thus,

$$\begin{aligned}
\|\widehat{b}_{h,\mathrm{h}} - b_0\|_{f,1,\mathrm{r}}^2 &\leqslant \frac{8}{m^2}\|\widehat{bf}_h - b_0 f\|_f^2 \\
&\quad + \frac{8}{m^2}\|(f - \widehat{f}_{\mathrm{h}})b_0\|_{f,1,\mathrm{r}}^2 + 2\|b_0\mathbf{1}_{|f(\cdot)-\widehat{f}_{\mathrm{h}}(\cdot)|>\frac{m}{2}}\|_{f,1,\mathrm{r}}^2 \\
&\leqslant \frac{8}{m^2}\int_{-\infty}^{\infty}(\widehat{bf}_h - b_0 f)(x)^2 f(x)dx \\
&\quad + \frac{8}{m^2}\int_{1}^{\mathrm{r}}(f(x) - \widehat{f}_{\mathrm{h}}(x))^2 b_0(x)^2 f(x)dx \\
&\quad + 2\int_{1}^{\mathrm{r}} b_0(x)^2 f(x)\mathbf{1}_{|f(x)-\widehat{f}_{\mathrm{h}}(x)|>\frac{m}{2}}dx.
\end{aligned}$$

Since f has a sub-Gaussian tail by Assumption 5.1, and since b_0 has at most linear growth because it is Lipschitz continuous from $\mathbb{R}$ into itself, $b_0^2 f$ is bounded on $\mathbb{R}$. So,

$$\begin{aligned}
\|\widehat{b}_{h,\mathrm{h}} - b_0\|_{f,1,\mathrm{r}}^2 &\leqslant \frac{8\|f\|_\infty}{m^2}\|\widehat{bf}_h - b_0 f\|^2 \\
&\quad + \frac{8\|b_0^2 f\|_\infty}{m^2}\|\widehat{f}_{\mathrm{h}} - f\|^2 + 2\|b_0^2 f\|_\infty\int_{-\infty}^{\infty}\mathbf{1}_{|f(x)-\widehat{f}_{\mathrm{h}}(x)|>\frac{m}{2}}dx.
\end{aligned}$$

Therefore, by Markov's inequality,

$$\begin{aligned}
\mathbb{E}(\|\widehat{b}_{h,\mathrm{h}} - b_0\|_{f,1,\mathrm{r}}^2) &\leqslant \frac{8\|f\|_\infty}{m^2}\mathbb{E}(\|\widehat{bf}_h - b_0 f\|^2) + \frac{8\|b_0^2 f\|_\infty}{m^2}\mathbb{E}(\|\widehat{f}_{\mathrm{h}} - f\|^2) \\
&\quad + \frac{8\|b_0^2 f\|_\infty}{m^2}\int_{-\infty}^{\infty}\mathbb{E}((f(x) - \widehat{f}_{\mathrm{h}}(x))^2)dx \\
&\leqslant \frac{8(\|f\|_\infty \vee \|b_0^2 f\|_\infty)}{m^2}(\mathbb{E}(\|\widehat{bf}_h - b_0 f\|^2) + 2\mathbb{E}(\|\widehat{f}_{\mathrm{h}} - f\|^2)).
\end{aligned}$$

Propositions 5.2 and 5.3 allow to conclude. □

Let us conclude this section with some remarks about Proposition 5.4:

- Proposition 5.4 says that the risk of $\widehat{b}_{h,\mathrm{h}}$ is controlled by the sum of those of its numerator and of its denominator, up to a multiplicative constant. If K fulfills Assumption 5.3 with $\upsilon = \beta \vee \gamma$, and if $b_0 f$ satisfies (5.1), then the risk bound on $\widehat{b}_{h,\mathrm{h}}$ is of order

$$h^{2\gamma} + \mathrm{h}^{2\beta} + \frac{1}{Nh} + \frac{1}{N\mathrm{h}},$$

and the bias-variance tradeoff is reached when h (resp. $\mathtt{h}$) is of order $N^{-1/(2\gamma+1)}$ (resp. $N^{-1/(2\beta+1)}$), leading to the rate

$$N^{-2\left(\left(\frac{\gamma}{2\gamma+1}\right)\wedge\left(\frac{\beta}{2\beta+1}\right)\right)} = N^{-\frac{2(\beta\wedge\gamma)}{2(\beta\wedge\gamma)+1}}.$$

- Note that to consider the 2-bandwidths estimator is crucial to extend the PCO (penalized comparison to overfitting) method to our framework in the spirit of Comte and Marie [4] (see Sect. 5.3). However, by taking $h = \mathtt{h}$ of order $N^{-1/(2(\beta\wedge\gamma)+1)}$, the bias-variance tradeoff is reached by the 1-bandwidth (truncated) Nadaraya-Watson estimator with the same rate.
- If $h = \mathtt{h}$, then the variance term in the risk bound in Proposition 5.4 is comparable to the variance term in the risk bound on the least squares projection estimator (see Theorem 3.1). Indeed, for a m-dimensional projection space, the variance term in the risk bound of Theorem 3.1 is of order m/N which is comparable to $1/(Nh)$. The rate of convergence of the projection least squares estimator depends on the regularity space associated to the projection basis but, as in the nonparametric regression framework, not on the regularity of f.
- Thanks to Menozzi et al. [2], Theorem 1.2, there exists $m \in (0, 1]$ such that $f(\cdot) > m$ on $[\mathtt{l}, \mathtt{r}]$. However, the limitation of our Proposition 5.4 is that m is unknown in practice. Then, it needs to be estimated, for instance by

$$\widehat{m}_{\mathtt{h}} = \min\{\widehat{f}_{\mathtt{h}}(x); x \in [\mathtt{l}, \mathtt{r}]\}.$$

A more naive way to solve this difficulty in practice is to take

$$m = m_N = \mathfrak{c}N^{-\frac{\varepsilon}{2}\cdot\frac{2(\beta\wedge\gamma)}{2(\beta\wedge\gamma)+1}} \xrightarrow[N\to\infty]{} 0,$$

where $\mathfrak{c} > 0$ is a fixed constant, and $\varepsilon \in (0, 1)$ is chosen as close as possible to 0. Under Assumption 5.1, by Proposition 5.1,

$$\exists N_0 \in \mathbb{N} : \forall N > N_0, \forall x \in [\mathtt{l}, \mathtt{r}], f(x) > m_N.$$

So, by Proposition 5.4, when h (resp. $\mathtt{h}$) is of order $N^{-1/(2\gamma+1)}$ (resp. $N^{-1/(2\beta+1)}$), $\widehat{b}_{h,\mathtt{h}}$ converges with the slightly degraded rate

$$N^{-(1-\varepsilon)\frac{2(\beta\wedge\gamma)}{2(\beta\wedge\gamma)+1}}.$$

5.3 Bandwidth Selection via the PCO Method

Consider

$$\widehat{s}_{h,\mathrm{v}}(x) = \frac{1}{N}\sum_{i=1}^{N}\Phi_{h,\mathrm{v}}^{i}(x);\ x \in \mathbb{R},$$

where $\mathrm{v} = (\lambda, \mu) \in \mathbb{R}^2$, $h > 0$ and, for any $i \in \{1, \dots, N\}$,

$$\Phi_{h,\mathrm{v}}^{i}(\cdot) = \frac{1}{T - t_0}\int_{t_0}^{T} K_h(X_s^i - \cdot)dY_s^i$$

with

$$\begin{aligned} dY_t^i &= \lambda dt + \mu dX_t^i \\ &= \underbrace{(\lambda + \mu b_0)}_{=:b_{\mathrm{v}}}(X_t^i)dt + \mu\sigma(X_t^i)dW_t^i \quad \text{and} \quad Y_0^i = \mu x_0. \end{aligned}$$

This is an estimator of the function $s = b_{\mathrm{v}} f$. In the sequel, for the sake of simplicity, $\widehat{s}_{h,\mathrm{v}}$ and $\Phi_{h,\mathrm{v}}^i$ are almost always denoted by $\widehat{s}_h$ and Φ_h^i respectively. An extension of the PCO method (see Sect. 2.1.2) to the selection of the bandwidth h of $\widehat{s}_h$ is given by

$$\widehat{h} = \arg\min_{h\in\mathcal{H}_N}\{\|\widehat{s}_h - \widehat{s}_{h_0}\|_{\delta}^2 + \mathrm{pen}(h)\} \tag{5.2}$$

where h_0 is an overfitting bandwidth belonging to $[N^{-1/3}, 1]$, $\mathcal{H}_N$ is a finite subset of $[h_0, 1]$, δ is a kernel and

$$\mathrm{pen}(h) = \frac{2}{N^2}\sum_{i=1}^{N}\langle\Phi_h^i(\cdot), \Phi_{h_0}^i(\cdot)\rangle_{\delta};\ \forall h \in \mathcal{H}_N.$$

This section deals with risk bounds on the PCO estimator $\widehat{s}_{\widehat{h}}$, and on the adaptive 2-bandwidths Nadaraya-Watson estimator

$$\widehat{b}_{\widehat{h},\widehat{\mathfrak{h}}}(x) = \frac{\widehat{bf}_{\widehat{h}}(x)}{\widehat{f}_{\widehat{\mathfrak{h}}}(x)}\mathbf{1}_{\widehat{f}_{\widehat{\mathfrak{h}}}(x) > \frac{m}{2}};\ x \in [\mathtt{l}, \mathtt{r}],$$

where $\mathtt{l}, \mathtt{r} \in \mathbb{R}$ satisfy $\mathtt{l} < \mathtt{r}$, and $\widehat{h}$ (resp. $\widehat{\mathfrak{h}}$) is defined by (5.2) with $\mathrm{v} = (0, 1)$ (resp. $\mathrm{v} = (1, 0)$).

Remark. The condition $h_0 \in [N^{-1/3}, 1]$ (instead of $h_0 \in [N^{-1}, 1]$) is annoying, but not that much because if, for instance, $b_0 f$ satisfies (5.1) with $\gamma = \beta \geqslant 2$, then the (unknown) bandwidth h_* for which the estimator of $b_0 f$ reaches the bias-variance tradeoff possibly belongs to $\mathcal{H}_N$. Indeed, there exists an unknown constant $\mathfrak{c}_* > 0$

such that $h_* = \mathfrak{c}_* N^{-1/(2\beta+1)}$, and then

$$\frac{1}{Nh_*^3} = \frac{1}{\mathfrak{c}_*^3} N^{\frac{2}{2\beta+1}(1-\beta)} \leqslant 1 \quad \text{for } N \text{ large enough.}$$

5.3.1 *Brief Reminder: Bernstein's Inequality and a Concentration Inequality for U-Statistics*

Consider a separable Banach space $\mathbb{X}$, a $\mathbb{X}$-valued random variable ξ, n independent copies $\xi_1, \dots, \xi_n$ of ξ with $n \in \mathbb{N}^*$, a bilinear form $\beta : \mathbb{X}^2 \to \mathbb{R}$, and the associated quadratic form $q : \mathbb{X} \to \mathbb{R}$. In order to establish $\mathbb{L}^2$-risk bounds on adaptive versions of functional estimators as $\widehat{f}_h$ and $\widehat{bf}_h$,

$$\mathbb{P}(|q(\overline{\xi}_n)| \leqslant M);\ M > 0 \quad \text{needs to be well-controlled.}$$

To that purpose, note that

$$q(\overline{\xi}_n) = \frac{U_n + V_n}{n^2},$$

where

$$U_n = \sum_{i \neq k} \beta(\xi_i, \xi_k) \quad \text{and} \quad V_n = \sum_{i=1}^{n} q(\xi_i).$$

Then, V_n is controlled thanks to Bernstein's inequality (see Theorem 5.1), and U_n is controlled thanks to a concentration inequality for U-statistics (see Theorem 5.2).

Theorem 5.1 (Weak Bernstein's inequality) *Let g be a bounded measurable map from $\mathbb{X}$ into $\mathbb{R}$, and consider*

$$v_n = \frac{1}{n} \sum_{i=1}^{n} (g(\xi_i) - \mathbb{E}(g(\xi_i))).$$

For every $\alpha > 0$, with probability larger than $1 - 2e^{-\alpha}$,

$$|v_n| \leqslant \sqrt{\frac{2\mathfrak{v}\alpha}{n}} + \frac{\mathfrak{c}\alpha}{n},$$

where

$$\mathfrak{c} = \frac{\|g\|_\infty}{3} \quad \textit{and} \quad \mathfrak{v} = \mathbb{E}(g(\xi)^2).$$

See Massart [5], Proposition 2.9 and Inequality (2.23) for a proof.

Theorem 5.2 *For every $i, k \in \{1, \dots, n\}$ with $i \neq k$, let $g_{i,k}$ be a bounded symmetric measurable map from $\mathbb{X}^2$ into $\mathbb{R}$ such that $g_{i,k} = g_{k,i}$ and*

$$\mathbb{E}(g_{i,k}(z, \xi)) = 0 \quad dz\text{-a.e.}$$

Consider the totally degenerate second order U-statistic

$$u_n = \sum_{i \neq k} g_{i,k}(\xi_i, \xi_k).$$

There exists a universal constant $\mathfrak{m} > 0$ such that, for every $\alpha > 0$, with probability larger than $1 - 5.4e^{-\alpha}$,

$$|u_n| \leqslant \mathfrak{m}(\mathfrak{c}_n \alpha^{\frac{1}{2}} + \mathfrak{d}_n \alpha + \mathfrak{b}_n \alpha^{\frac{3}{2}} + \mathfrak{a}_n \alpha^2),$$

where

$$\mathfrak{a}_n = \sup_{i,k \in \{1,\dots,n\}} \left\{ \sup_{x,z \in \mathbb{X}} |g_{i,k}(z, x)| \right\},$$

$$\mathfrak{b}_n^2 = \max \left\{ \sup_{i,z} \sum_{k=1}^{i-1} \mathbb{E}(g_{i,k}(z, \xi_k)^2); \sup_{k,x} \sum_{i=k+1}^{n} \mathbb{E}(g_{i,k}(\xi_i, x)^2) \right\},$$

$$\mathfrak{c}_n^2 = \sum_{i \neq k} \mathbb{E}(g_{i,k}(\xi_i, \xi_k)^2) \quad \text{and}$$

$$\mathfrak{d}_n = \sup_{(a,b) \in \mathcal{A}} \sum_{i<k} \mathbb{E}(a_i(\xi_i) b_k(\xi_k) g_{i,k}(\xi_i, \xi_k))$$

with

$$\mathcal{A} = \left\{ (a, b) : \sum_{i=1}^{n-1} \mathbb{E}(a_i(\xi_i)^2) \leqslant 1 \text{ and } \sum_{k=2}^{n} \mathbb{E}(b_k(\xi_k)^2) \leqslant 1 \right\}.$$

See Giné and Nickl [6], Theorem 3.4.8 for a proof. Finally, for every random variable R, the following proposition provides a bound on $\mathbb{E}(R)$ from those on $\mathbb{P}(R \leqslant M)$ $(M > 0)$.

Proposition 5.5 *Let R be a random variable, and assume that there exist $r, c > 0$ and $p \geqslant 1$ such that, for every $\alpha \in \mathbb{R}_+$,*

$$\mathbb{P}\left(R \leqslant \frac{\alpha^p}{r} \right) \geqslant 1 - ce^{-\alpha}.$$

Then,

$$\mathbb{E}(R) \leqslant \frac{2^{p+1}\log(c)^p}{r} + \frac{\mathfrak{c}_p}{r} \quad \textit{with} \quad \mathfrak{c}_p = \int_0^\infty \exp\left(-\frac{1}{2}\beta^{\frac{1}{p}}\right)d\beta < \infty.$$

Proof Consider $A > 0$. First, by the Fubini-Tonelli theorem,

$$\begin{aligned}\mathbb{E}(R) &= \mathbb{E}(R\mathbf{1}_{R\leqslant A}) + \mathbb{E}((R-A)\mathbf{1}_{R>A}) + A\mathbb{P}(R > A)\\ &\leqslant 2A + \mathbb{E}\left(\mathbf{1}_{R>A}\int_A^\infty \mathbf{1}_{R>x}dx\right) \leqslant 2A + \int_A^\infty \mathbb{P}(R > x)dx.\end{aligned}$$

Now, by the change of variable formula,

$$\begin{aligned}\int_A^\infty \mathbb{P}(R > x)dx &= \frac{1}{r}\int_{rA}^\infty \mathbb{P}\left(R > \frac{\beta}{r}\right)d\beta\\ &\leqslant \frac{c}{r}\int_{rA}^\infty e^{-\beta^{1/p}}d\beta \leqslant \frac{\mathfrak{c}_p c}{r}\exp\left(-\frac{1}{2}(rA)^{\frac{1}{p}}\right).\end{aligned}$$

Then, for $A = \log(c^2)^p/r$,

$$\begin{aligned}\mathbb{E}(R) &\leqslant \frac{2\log(c^2)^p}{r} + \frac{\mathfrak{c}_p c}{r}\exp\left(-\frac{1}{2}\log(c^2)\right)\\ &= \frac{2^{p+1}\log(c)^p}{r} + \frac{\mathfrak{c}_p}{r}.\end{aligned}$$

□

For instance, with the notations of Theorem 5.1, the random variable

$$R = \frac{1}{\mathfrak{c}}\left(|v_n| - \sqrt{\frac{2\mathfrak{v}\alpha}{n}}\right)$$

satisfies

$$\mathbb{P}\left(R \leqslant \frac{\alpha}{n}\right) \geqslant 1 - 2e^{-\alpha};\ \forall\alpha \in \mathbb{R}_+,$$

and then

$$\mathbb{E}(R) \leqslant \frac{4\log(2)}{n} + \frac{\mathfrak{c}_1}{n} \quad \text{by Proposition 5.5.}$$

Therefore,

$$\mathbb{E}(|v_n|) \leqslant \sqrt{\frac{2\mathfrak{v}\alpha}{n}} + \frac{\bar{\mathfrak{c}}}{n} \quad \text{with} \quad \bar{\mathfrak{c}} = \mathfrak{c}(4\log(2) + \mathfrak{c}_1).$$

5.3.2 Risk Bound on the Adaptive 2-Bandwidths Nadaraya-Watson Estimator

This section deals with a risk bound on the PCO estimator of s, and then with a risk bound on the (PCO) adaptive 2-bandwidths Nadaraya-Watson estimator of b_0. Throughout this section, the kernels K and δ fulfill the following assumption.

Assumption 5.4 The kernels K and δ are continuously differentiable, $K' \in \mathbb{L}^2(\mathbb{R}, dx)$ and δ is nonnegative.

Note that the standard normal density function is a nice example of kernel fulfilling all the conditions on both K and δ in Assumption 5.4.

First of all, consider $h \in (0, 1]$, and let us show that there exists a kernel type map Φ_h from $C^0([0, T]) \times \mathbb{R}$ into $\mathbb{R}$ such that, for any $i \in \{1, \dots, N\}$, $\Phi_h^i(\cdot) = \Phi_h(X^i, \cdot)$. By the Itô formula,

$$\int_{t_0}^{T} K_h(X_s^i - \cdot)dX_s^i = \mathrm{K}\left(\frac{X_T^i - \cdot}{h}\right) - \mathrm{K}\left(\frac{X_{t_0}^i - \cdot}{h}\right)$$
$$- \frac{1}{2h^2}\int_{t_0}^{T} K'\left(\frac{X_s^i - \cdot}{h}\right)\sigma(X_s^i)^2 ds \quad \text{with} \quad \mathrm{K}' = K.$$

Then,

$$(T - t_0)\Phi_h^i(\cdot) = \lambda\int_{t_0}^{T} K_h(X_s^i - \cdot)ds + \mu\int_{t_0}^{T} K_h(X_s^i - \cdot)dX_s^i$$
$$= \mu\mathrm{K}\left(\frac{X_T^i - \cdot}{h}\right) - \mu\mathrm{K}\left(\frac{X_{t_0}^i - \cdot}{h}\right) + \int_{t_0}^{T} K_{h,\mathrm{v},\sigma}(X_s^i, \cdot)ds$$

where, for every $z \in \mathbb{R}$,

$$K_{h,\mathrm{v},\sigma}(z, \cdot) = \lambda K_h(z - \cdot) + \mu\overline{K}_h(z - \cdot)\sigma(z)^2$$
$$\text{with} \quad \overline{K}_h(\cdot) = -\frac{1}{2h^2}K'\left(\frac{\cdot}{h}\right).$$

So, $\Phi_h^i(\cdot) = \Phi_h(X^i, \cdot)$ with

$$\Phi_h(\varphi, x) := \frac{1}{T - t_0}\left(\mu\mathrm{K}\left(\frac{\varphi(T) - x}{h}\right) - \mu\mathrm{K}\left(\frac{\varphi(t_0) - x}{h}\right) + \int_{t_0}^{T} K_{h,\mathrm{v},\sigma}(\varphi(s), x)ds\right).$$

In order to establish a suitable risk bound on the PCO estimator of s, the following proposition shows that

$$\mathcal{K}_N = \{(\varphi, x) \mapsto \Phi_h(\varphi, x); h \in \mathcal{H}_N\}$$

satisfies properties close to those of a kernels set in the nonparametric regression framework (see Assumptions 2.1 and 2.3).

Proposition 5.6 *Under Assumptions 5.1, 5.2 and 5.4, there exists a constant* $\mathfrak{m}_\Phi > 0$ *such that*

1. *For every* $h \in (0, 1]$ *and* $\varphi \in C^0([0, T])$,

$$|\Phi_h(\varphi, \cdot)| \leqslant \frac{\mathfrak{m}_\Phi}{h^2} \quad \text{and} \quad \|\Phi_h(\varphi, \cdot)\|_\delta^2 \leqslant \frac{\mathfrak{m}_\Phi}{h^3}.$$

2. *For every* $h,\quad h \in (0, 1]$,

$$\mathbb{E}(\langle \Phi_h(X^1, \cdot), \Phi_h(X^2, \cdot)\rangle_\delta^2) \leqslant \mathfrak{m}_\Phi \overline{s}_h,$$

where

$$\overline{s}_h = \mathbb{E}(\|\Phi_h(X, \cdot)\|_\delta^2).$$

3. *For every* $h \in (0, 1]$ *and* $\varphi \in \mathbb{L}^2(\mathbb{R}, dx)$,

$$\mathbb{E}(\langle \Phi_h(X, \cdot), \varphi\rangle_\delta^2) \leqslant \mathfrak{m}_\Phi \|\varphi\|_\delta^2.$$

4. *For every* $h,\quad h \in (0, 1]$,

$$|\langle \Phi_h(X, \cdot), s_h\rangle_\delta| \leqslant \mathfrak{m}_\Phi \quad a.s,$$

where

$$s_h(\cdot) = \mathbb{E}(\widehat{s}_h(\cdot)) = (K_h * s)(\cdot).$$

Proof First of all, note that

- Since b_0' and σ are bounded, and by Assumption 5.1, the functions $\sigma^2 f$, $b^2{}_{\mathrm{v}} f$, s and s' are bounded and integrable.
- Since K is a kernel, and by Assumption 5.4, K, K' and K are bounded, $\|K\|_1 < \infty$ and $\|K'\| < \infty$.
- Since δ is a kernel, and by Assumption 5.4, δ and δ' are bounded.

Thanks to these properties of σ, b_{v}, s, s', K, K', K, δ and δ', let us prove Proposition 5.6.

1. On the one hand, by the definition of the map $(h, \varphi, x) \mapsto \Phi_h(\varphi, x)$, for every $h \in (0, 1]$ and $\varphi \in C^0([0, T])$,

$$|\Phi_h(\varphi, \cdot)| \leqslant \frac{\lambda\|K\|_\infty}{h} + \mu\left(\frac{2\|\ \mathrm{K}\|_\infty}{T - t_0} + \frac{\|\sigma\|_\infty^2\|K'\|_\infty}{2h^2}\right).$$

On the other hand, for any $h \in (0, 1]$ and $z \in \mathbb{R}$,

$$\|K_{h,\mathrm{v},\sigma}(z,\cdot)\|_\delta^2 \leqslant 2\lambda^2\|K_h(z-\cdot)\|_\delta^2 + 2\mu^2\|\sigma\|_\infty^4\|\overline{K}_h(z-\cdot)\|_\delta^2$$
$$\leqslant \frac{2\lambda^2\|\delta\|_\infty\|K\|^2}{h} + \frac{\mu^2\|\delta\|_\infty\|\sigma\|_\infty^4\|K'\|^2}{2h^3}.$$

So, for every $\varphi \in C^0([0,T])$,

$$\|\Phi_h(\varphi,\cdot)\|_\delta^2 \leqslant \frac{2\mu^2}{(T-t_0)^2}\int_{-\infty}^{\infty}\left(\mathrm{K}\left(\frac{\varphi(T)-x}{h}\right) - \mathrm{K}\left(\frac{\varphi(t_0)-x}{h}\right)\right)^2\delta(x)dx$$
$$+ \frac{2}{(T-t_0)^2}\left\|\int_{t_0}^{T} K_{h,\mathrm{v},\sigma}(\varphi(s),\cdot)ds\right\|_\delta^2$$
$$\leqslant \frac{8\mu^2\|\mathrm{K}\|_\infty^2}{(T-t_0)^2} + \frac{2}{T-t_0}\int_{t_0}^{T}\|K_{h,\mathrm{v},\sigma}(\varphi(s),\cdot)\|_\delta^2 ds$$
$$\leqslant \frac{4\lambda^2\|\delta\|_\infty\|K\|^2}{h} + \mu^2\left(\frac{8\|\mathrm{K}\|_\infty^2}{(T-t_0)^2} + \frac{\|\delta\|_\infty\|\sigma\|_\infty^4\|K'\|^2}{h^3}\right).$$

2. For any $h, \mathrm{h} \in (0, 1]$,

$$\mathbb{E}(\langle\Phi_h(X^1,\cdot),\Phi_{\mathrm{h}}(X^2,\cdot)\rangle_\delta^2)$$
$$= \frac{1}{(T-t_0)^2}\mathbb{E}\left[\left(\int_{-\infty}^{\infty}\left(\int_{t_0}^{T} K_h(X_s^1-x)dY_s^1\right)\Phi_{\mathrm{h}}(X^2,x)\delta(x)dx\right)^2\right]$$
$$\leqslant \frac{2}{(T-t_0)^2}(\mathbb{E}(A_{h,\mathrm{h}}^2) + \mathbb{E}(B_{h,\mathrm{h}}^2)),$$

where

$$A_{h,\mathrm{h}} = \mu\int_{-\infty}^{\infty}\left(\int_{t_0}^{T} K_h(X_s^1-x)\sigma(X_s^1)dW_s^1\right)\Phi_{\mathrm{h}}(X^2,x)\delta(x)dx$$
$$\text{and}\quad B_{h,\mathrm{h}} = \int_{-\infty}^{\infty}\left(\int_{t_0}^{T} K_h(X_s^1-x)b_{\mathrm{v}}(X_s^1)ds\right)\Phi_{\mathrm{h}}(X^2,x)\delta(x)dx.$$

On the one hand, since (X^1, W^1) and Y^2 are independent processes,

$$\mathbb{E}(A_{h,\mathrm{h}}^2) = \mu^2\int_{-\infty}^{\infty}\int_{-\infty}^{\infty} A_h(x,y)\mathbb{E}(\Phi_{\mathrm{h}}(X,x)\Phi_{\mathrm{h}}(X,y))\delta(x)\delta(y)dxdy$$

where, for every $x, y \in \mathbb{R}$,

$$A_h(x,y) = \mathbb{E}\left(\left(\int_{t_0}^{T} K_h(X_s-x)\sigma(X_s)dW_s\right)\left(\int_{t_0}^{T} K_h(X_s-y)\sigma(X_s)dW_s\right)\right).$$

By the isometry property of Itô's integral, for every $x, y \in \mathbb{R}$,

$$A_h(x, y) = \int_{t_0}^{T} \mathbb{E}(K_h(X_s - x)K_h(X_s - y)\sigma(X_s)^2)ds$$
$$= (T - t_0) \int_{-\infty}^{\infty} K_h(z - x)K_h(z - y)\sigma(z)^2 f(z)dz,$$

and then

$$\mathbb{E}(A_{h,\mathrm{h}}^2) = \mu^2(T - t_0) \int_{-\infty}^{\infty}\int_{-\infty}^{\infty}\int_{-\infty}^{\infty} K_h(z - x)K_h(z - y)\sigma(z)^2 f(z) \times \mathbb{E}(\Phi_{\mathrm{h}}(X, x)\Phi_{\mathrm{h}}(X, y))\delta(x)\delta(y)dxdydz$$
$$= \mu^2(T - t_0) \int_{-\infty}^{\infty} \sigma(z)^2 f(z)\mathbb{E}\left[\left(\int_{-\infty}^{\infty} K_h(z - x)\delta(x)\Phi_{\mathrm{h}}(X, x)dx\right)^2\right]dz.$$

Moreover, since $x \mapsto |K_h(z - x)|/\|K\|_1$ is a density function for any $z \in \mathbb{R}$,

$$\mathbb{E}\left[\left(\int_{-\infty}^{\infty} K_h(z - x)\delta(x)\Phi_{\mathrm{h}}(X, x)dx\right)^2\right] \leqslant \|K\|_1 \int_{-\infty}^{\infty} |K_h(z - x)|\delta(x)^2\mathbb{E}(\Phi_{\mathrm{h}}(X, x)^2)dx.$$

So,

$$\mathbb{E}(A_{h,\mathrm{h}}^2) \leqslant \mu^2(T - t_0)\|K\|_1 \times \int_{-\infty}^{\infty} \sigma(z)^2 f(z) \int_{-\infty}^{\infty} |K_h(z - x)|\delta(x)^2\mathbb{E}(\Phi_{\mathrm{h}}(X, x)^2)dxdz$$
$$\leqslant \mu^2(T - t_0)\|\sigma^2 f\|_\infty \|K\|_1^2 \|\delta\|_\infty \underbrace{\int_{-\infty}^{\infty} \delta(x)\mathbb{E}(\Phi_{\mathrm{h}}(X, x)^2)dx}_{=\overline{s}_{\mathrm{h}}}.$$

On the other hand, since $x \mapsto |K_h(X_t(\omega) - x)|/\|K\|_1$ is a density function, and since X^1 and X^2 are independent processes,

$$\mathbb{E}(B_{h,\mathrm{h}}^2) \leqslant (T - t_0)\|K\|_1 \times \int_{t_0}^{T}\int_{-\infty}^{\infty} \mathbb{E}(|K_h(X_s - x)|b_{\mathrm{v}}(X_s)^2)\delta(x)^2\mathbb{E}(\Phi_{\mathrm{h}}(X, x)^2)dxds$$
$$= (T - t_0)^2\|K\|_1 \times \int_{-\infty}^{\infty}\left(\int_{-\infty}^{\infty} |K_h(z - x)|b_{\mathrm{v}}(z)^2 f(z)dz\right)\delta(x)^2\mathbb{E}(\Phi_{\mathrm{h}}(X, x)^2)dx$$

$$\leqslant (T - t_0)^2 \|b^2{}_{\mathrm{v}} f\|_\infty \|K\|_1^2 \|\delta\|_\infty \underbrace{\int_{-\infty}^{\infty} \delta(x)\mathbb{E}(\Phi_{\mathrm{h}}(X, x)^2)dx}_{=\overline{s}_{\mathrm{h}}}.$$

Therefore,

$$\mathbb{E}(\langle \Phi_h(X^1, \cdot), \Phi_{\mathrm{h}}(X^2, \cdot)\rangle_\delta^2) \leqslant \frac{2\|K\|_1^2 \|\delta\|_\infty}{T - t_0}(\mu^2 \|\sigma^2 f\|_\infty + (T - t_0)\|b^2{}_{\mathrm{v}} f\|_\infty)\overline{s}_{\mathrm{h}}.$$

3. For any $h \in (0, 1]$ and $\varphi \in \mathbb{L}^2(\mathbb{R}, dx)$,

$$\mathbb{E}(\langle \Phi_h(X, \cdot), \varphi\rangle_\delta^2) = \frac{1}{(T - t_0)^2}\mathbb{E}\left[\left(\int_{-\infty}^{\infty} \varphi(x)\delta(x) \int_{t_0}^{T} K_h(X_s - x)dY_s dx\right)^2\right]$$
$$\leqslant \frac{2}{(T - t_0)^2}(\mathbb{E}(C_h^2) + \mathbb{E}(D_h^2)),$$

where

$$C_h = \mu \int_{-\infty}^{\infty} \varphi(x)\delta(x) \int_{t_0}^{T} K_h(X_s - x)\sigma(X_s)dW_s dx$$
$$\text{and} \quad D_h = \int_{-\infty}^{\infty} \varphi(x)\delta(x) \int_{t_0}^{T} K_h(X_s - x)b_{\ \mathrm{v}}(X_s)ds dx.$$

On the one hand, by the isometry property of Itô's integral,

$$\mathbb{E}(C_h^2) = \mu^2 \int_{-\infty}^{\infty}\int_{-\infty}^{\infty} \varphi(x)\varphi(y)\delta(x)\delta(y)$$
$$\times \int_{t_0}^{T} \mathbb{E}(K_h(X_s - x)K_h(X_s - y)\sigma(X_s)^2)ds dx dy$$
$$= \mu^2(T - t_0)\int_{-\infty}^{\infty}\int_{-\infty}^{\infty} \varphi(x)\varphi(y)\delta(x)\delta(y)$$
$$\times \int_{-\infty}^{\infty} K_h(z - x)K_h(z - y)\sigma(z)^2 f(z)dz dx dy$$
$$= \mu^2(T - t_0)\underbrace{\int_{-\infty}^{\infty} (K_h * (\varphi\delta))(z)^2\sigma(z)^2 f(z)dz}_{\leqslant \|\sigma^2 f\|_\infty \|K\|_1^2 \|\delta\|_\infty \|\varphi\|_\delta^2}.$$

On the other hand,

$$\mathbb{E}(D_h^2) \leqslant (T - t_0)\int_{t_0}^{T} \mathbb{E}(b_{\ \mathrm{v}}(X_s)^2(K_h * (\varphi\delta))(X_s)^2)ds$$

$$\leqslant (T-t_0)^2 \underbrace{\int_{-\infty}^{\infty}(K_h*(\varphi\delta))(z)^2 b_{\mathrm{v}}(z)^2 f(z)dz}_{\leqslant \|b^2{}_{\mathrm{v}}f\|_\infty \|K\|_1^2\|\delta\|_\infty\|\varphi\|_\delta^2}.$$

Therefore,

$$\mathbb{E}(\langle\Phi_h(X,\cdot),\varphi\rangle_\delta^2) \leqslant \frac{2}{T-t_0}(\mu^2\|\sigma^2 f\|_\infty + (T-t_0)\|b^2{}_{\mathrm{v}}f\|_\infty)\|K\|_1^2\|\delta\|_\infty\|\varphi\|_\delta^2.$$

4. Since the map $(t,\omega,x)\mapsto K_h(X_t(\omega)-x)s_{\mathrm{h}}(x)$ is measurable and bounded for any $h,\mathrm{h}\in(0,1]$, and since

$$A\mapsto \int_A \delta(x)dx \quad \text{is a finite measure,}$$

by the stochastic Fubini theorem,

$$\begin{aligned}(T-t_0)\langle\Phi_h(X,\cdot),s_{\mathrm{h}}\rangle_\delta &= (T-t_0)\int_{-\infty}^{\infty}\Phi_h(X,x)s_{\mathrm{h}}(x)\delta(x)dx\\ &= \int_{t_0}^{T}\int_{-\infty}^{\infty}K_h(X_s-x)s_{\mathrm{h}}(x)\delta(x)dxdY_s \quad \text{a.s.}\\ &= \lambda\int_{t_0}^{T}(K_h*(s_{\mathrm{h}}\delta))(X_s)ds + \mu\int_{t_0}^{T}(K_h*(s_{\mathrm{h}}\delta))(X_s)dX_s.\end{aligned}$$

First,

$$\begin{aligned}\frac{1}{T-t_0}\left|\int_{t_0}^{T}(K_h*(s_{\mathrm{h}}\delta))(X_s)ds\right| &\leqslant \|K_h*(s_{\mathrm{h}}\delta)\|_\infty\\ &\leqslant \|K\|_1\|(K_{\mathrm{h}}*s)\delta\|_\infty \leqslant \|K\|_1^2\|s\|_\infty\|\delta\|_\infty.\end{aligned}$$

Now, by the Itô formula,

$$\int_{t_0}^{T}(K_h*(s_{\mathrm{h}}\delta))(X_s)dX_s = \Psi_{h,\mathrm{h}}(X_T)-\Psi_{h,\mathrm{h}}(X_{t_0}) - \frac{1}{2}\int_{t_0}^{T}\psi_{h,\mathrm{h}}'(X_s)\sigma(X_s)^2ds,$$

where

$$\psi_{h,\mathrm{h}} = K_h*(s_{\mathrm{h}}\delta) \quad\text{and}\quad \Psi_{h,\mathrm{h}}(\cdot) = \mathrm{K}\left(\frac{\cdot}{h}\right)*(s_{\mathrm{h}}\delta).$$

On the one hand,

$$\psi_{h,\mathrm{h}}' = K_h*(s_{\mathrm{h}}\delta)' = K_h*((K_{\mathrm{h}}*s)\delta') + K_h*((K_{\mathrm{h}}*s')\delta),$$

and then

$$\begin{aligned}\|\psi'_{h,\mathrm{h}}\|_\infty &\leqslant \|K_h\|_1\|K_{\mathrm{h}} * s\|_\infty\|\delta'\|_\infty + \|K_h\|_1\|K_{\mathrm{h}} * s'\|_\infty\|\delta\|_\infty\\ &\leqslant \|K\|_1^2\|s\|_\infty\|\delta'\|_\infty + \|K\|_1^2\|s'\|_\infty\|\delta\|_\infty.\end{aligned}$$

On the other hand,

$$\|\Psi_{h,\mathrm{h}}\|_\infty \leqslant \left\|\mathrm{K}\left(\frac{\cdot}{h}\right)\right\|_\infty \|(K_{\mathrm{h}} * s)\delta\|_1 \leqslant \|\mathrm{K}\|_\infty\|K\|_1\|s\|_1\|\delta\|_\infty.$$

Therefore,

$$\begin{aligned}\langle \Phi_h(X,\cdot), s_{\mathrm{h}}\rangle_\delta \leqslant \lambda\|K\|_1^2\|s\|_\infty\|\delta\|_\infty &+ \frac{2\mu}{T-t_0}\|\mathrm{K}\|_\infty\|K\|_1\|s\|_1\|\delta\|_\infty\\ &+ \frac{\mu}{2}\|\sigma\|_\infty^2\|K\|_1^2(\|s\|_\infty\|\delta'\|_\infty + \|s'\|_\infty\|\delta\|_\infty) \quad \text{a.s.}\end{aligned}$$

□

The following theorem, which is the main result of this section, provides a risk bound on $\widehat{s}_{\widehat{h}}$, where $\widehat{h}$ is defined by (5.2).

Theorem 5.3 *Under Assumptions 5.1, 5.2 and 5.4, there exist two deterministic constants* $\mathfrak{c}_{5.3}, \overline{\mathfrak{c}}_{5.3} > 0$, *not depending on* N, *such that for every* $\theta \in (0,1)$ *and* $\alpha > 0$, *with probability larger than* $1 - \overline{\mathfrak{c}}_{5.3}|\mathcal{H}_N|e^{-\alpha}$,

$$\|\widehat{s}_{\widehat{h}} - s\|_\delta^2 \leqslant (1+\theta)\min_{h\in\mathcal{H}_N}\|\widehat{s}_h - s\|_\delta^2 + \frac{\mathfrak{c}_{5.3}}{\theta}\left(\|s_{h_0} - s\|_\delta^2 + \frac{(1+\alpha)^3}{N}\right).$$

Proof For every $h, \mathrm{h} \in \mathcal{H}_N$, consider

$$U_{h,\mathrm{h}} = \sum_{i\neq k}\langle \Phi_h(X^i,\cdot) - s_h, \Phi_{\mathrm{h}}(X^k,\cdot) - s_{\mathrm{h}}\rangle_\delta,$$

$$V_h = \frac{1}{N}\sum_{i=1}^N\|\Phi_h(X^i,\cdot) - s_h\|_\delta^2 \quad\text{and}\quad W_{h,\mathrm{h}} = \langle \widehat{s}_h - s_h, s_{\mathrm{h}} - s\rangle_\delta.$$

The proof of Theorem 5.3 is dissected in four steps. Step 1 shows that, for any $h \in \mathcal{H}_N$,

$$\|\widehat{s}_{\widehat{h}} - s\|_\delta^2 \leqslant \|\widehat{s}_h - s\|_\delta^2 - \psi(h) + \psi(\widehat{h}),$$

where ψ is a map depending on U and W. Then, $\psi(h)$ and $\psi(\widehat{h})$ are controlled in Step 2 thanks to the bounds on U and W established in Sect. 5.3.2.1 (see Lemmas 5.1 and 5.3 respectively). Step 3 deals with a two-sided relationship between

$$\|\widehat{s}_h - s\|_\delta^2 \quad\text{and}\quad \|s_h - s\|_\delta^2;\ h \in \mathcal{H}_N,$$

as in Theorem 2.1 and the first step of the (sketch of) proof of Theorem 2.2, thanks to Lemmas 5.1, 5.2 (bound on V) and 5.3. The conclusion comes in Step 4. Finally, note that as in the nonparametric regression framework, the properties of

$$\mathcal{K}_N = \{(\varphi, x) \mapsto \Phi_h(\varphi, x); h \in \mathcal{H}_N\}$$

stated in Proposition 5.6 are crucial in the proof of Theorem 5.3.
Step 1. First,

$$\|\widehat{s}_{\widehat{h}} - s\|_\delta^2 = \|\widehat{s}_{\widehat{h}} - \widehat{s}_{h_0}\|_\delta^2 + \|\widehat{s}_{h_0} - s\|_\delta^2 + 2\langle \widehat{s}_{\widehat{h}} - \widehat{s}_{h_0}, \widehat{s}_{h_0} - s\rangle_\delta$$

and, for any $h \in \mathcal{H}_N$,

$$\begin{aligned}
\|\widehat{s}_{\widehat{h}} - \widehat{s}_{h_0}\|_\delta^2 &\leqslant \|\widehat{s}_h - \widehat{s}_{h_0}\|_\delta^2 + \text{pen}(h) - \text{pen}(\widehat{h}) \quad \text{by (5.2)}\\
&= \|\widehat{s}_h - s\|_\delta^2 + 2\langle \widehat{s}_h - s, s - \widehat{s}_{h_0}\rangle_\delta + \|s - \widehat{s}_{h_0}\|_\delta^2 + \text{pen}(h) - \text{pen}(\widehat{h})\\
&= \|\widehat{s}_h - s\|_\delta^2 + 2\langle \widehat{s}_h - \widehat{s}_{h_0}, s - \widehat{s}_{h_0}\rangle_\delta - \|s - \widehat{s}_{h_0}\|_\delta^2 + \text{pen}(h) - \text{pen}(\widehat{h}).
\end{aligned}$$

Then,

$$\begin{aligned}
\|\widehat{s}_{\widehat{h}} - s\|_\delta^2 &\leqslant \|\widehat{s}_h - s\|_\delta^2 + 2\langle \widehat{s}_h - \widehat{s}_{h_0}, s - \widehat{s}_{h_0}\rangle_\delta\\
&\qquad + \text{pen}(h) - \text{pen}(\widehat{h}) + 2\langle \widehat{s}_{\widehat{h}} - \widehat{s}_{h_0}, \widehat{s}_{h_0} - s\rangle_\delta\\
&= \|\widehat{s}_h - s\|_\delta^2 + \text{pen}(h) - \text{pen}(\widehat{h}) + 2\langle \widehat{s}_{\widehat{h}} - \widehat{s}_h, \widehat{s}_{h_0} - s\rangle_\delta\\
&= \|\widehat{s}_h - s\|_\delta^2 - \psi(h) + \psi(\widehat{h}) \qquad (5.3)
\end{aligned}$$

where

$$\psi(\cdot) = 2\langle \widehat{s}_{\cdot} - s, \widehat{s}_{h_0} - s\rangle_\delta - \text{pen}(\cdot).$$

Now, let us rewrite $\psi(\cdot)$ in terms of $U_{.,h_0}$, $W_{.,h_0}$ and $W_{h_0,.}$. For any $h \in \mathcal{H}_N$,

$$\begin{aligned}
\psi(h) &= 2\langle \widehat{s}_h - s_h + s_h - s, \widehat{s}_{h_0} - s_{h_0} + s_{h_0} - s\rangle_\delta - \text{pen}(h)\\
&= 2\langle \widehat{s}_h - s_h, \widehat{s}_{h_0} - s_{h_0}\rangle_\delta - \text{pen}(h)\\
&\qquad + 2\underbrace{\left(W_{h,h_0} + W_{h_0,h} + \langle s_h - s, s_{h_0} - s\rangle_\delta\right)}_{=:\psi_3(h)}\\
&= \frac{2U_{h,h_0}}{N^2} + 2\underbrace{\left(\frac{1}{N^2}\sum_{i=1}^{N}\langle \Phi_h(X^i, \cdot) - s_h, \Phi_{h_0}(X^i, \cdot) - s_{h_0}\rangle_\delta - \frac{\text{pen}(h)}{2}\right)}_{=:\psi_2(h)} + 2\psi_3(h)
\end{aligned}$$

and, by the definition of pen(h),

$$\psi_2(h) = -\frac{1}{N^2}\left(\sum_{i=1}^{N}\langle \Phi_h(X^i,\cdot), s_{h_0}\rangle_\delta + \sum_{i=1}^{N}\langle \Phi_{h_0}(X^i,\cdot), s_h\rangle_\delta\right) + \frac{1}{N}\langle s_h, s_{h_0}\rangle_\delta.$$

So,

$$\psi(h) = 2(\psi_1(h) + \psi_2(h) + \psi_3(h)) \quad \text{with} \quad \psi_1(h) = \frac{U_{h,h_0}}{N^2}.$$

Step 2. This step deals with suitable bounds on the ψ_j's.

- Consider $h \in \mathcal{H}_N$. By Lemma 5.1, for any $\alpha > 0$ and $\theta \in (0, 1)$, with probability larger than $1 - 5.4|\mathcal{H}_N|e^{-\alpha}$,

$$|\psi_1(h)| \leqslant \frac{\theta}{2N}\overline{s}_h + \frac{2\mathfrak{c}_{5.1}(1+\alpha)^3}{\theta N}$$

$$\text{and} \quad |\psi_1(\widehat{h})| \leqslant \frac{\theta}{2N}\overline{s}_{\widehat{h}} + \frac{2\mathfrak{c}_{5.1}(1+\alpha)^3}{\theta N}.$$

- For any $h, \mathrm{h} \in \mathcal{H}_N$, consider

$$\overline{\psi}_2(h, \mathrm{h}) = \frac{1}{N}\sum_{i=1}^{N}\langle \Phi_h(X^i,\cdot), s_{\mathrm{h}}\rangle_\delta.$$

By Proposition 5.6(4),

$$|\overline{\psi}_2(h, \mathrm{h})| \leqslant \mathfrak{m}_\Phi \quad \text{a.s.}$$

Moreover,

$$|\langle s_h, s_{h_0}\rangle_\delta| \leqslant \|\delta\|_\infty \|K_h * s\|_\infty \|K_{h_0} * s\|_1$$
$$\leqslant \|\delta\|_\infty \|K\|_1^2 \|s\|_\infty \|s\|_1.$$

Then, there exists a deterministic constant $\mathfrak{c}_1 > 0$, not depending on N, such that

$$|\psi_2(h)| \vee |\psi_2(\widehat{h})| \leqslant \sup_{\mathrm{h}\in\mathcal{H}_N} |\psi_2(\mathrm{h})| \leqslant \frac{\mathfrak{c}_1}{N} \quad \text{a.s.}$$

- Consider $h \in \mathcal{H}_N$. By Lemma 5.3 and Cauchy-Schwarz's inequality, with probability larger than $1 - |\mathcal{H}_N|e^{-\alpha}$,

$$|\psi_3(h)| \leqslant \frac{\theta}{4}(\|s_h - s\|_\delta^2 + \|s_{h_0} - s\|_\delta^2) + \frac{8\mathfrak{c}_{5.3}(1+\alpha)^2}{\theta N}$$
$$+ 2\left(\frac{1}{2}\right)^{\frac{1}{2}}\left(\frac{\theta}{2}\right)^{\frac{1}{2}}\|s_h - s\|_\delta \times \left(\frac{1}{2}\right)^{\frac{1}{2}}\left(\frac{2}{\theta}\right)^{\frac{1}{2}}\|s_{h_0} - s\|_\delta$$
$$\leqslant \frac{\theta}{2}\|s_h - s\|_\delta^2 + \left(\frac{\theta}{4} + \frac{1}{\theta}\right)\|s_{h_0} - s\|_\delta^2 + \frac{8\mathfrak{c}_{5.3}(1+\alpha)^2}{\theta N}$$

and

$$|\psi_3(\widehat{h})| \leqslant \frac{\theta}{2}\|s_{\widehat{h}} - s\|_\delta^2 + \left(\frac{\theta}{4} + \frac{1}{\theta}\right)\|s_{h_0} - s\|_\delta^2 + \frac{8\mathfrak{c}_{5.3}(1+\alpha)^2}{\theta N}.$$

Step 3. Let us establish that there exist two deterministic constants $\mathfrak{c}_2, \overline{\mathfrak{c}}_2 > 0$, not depending on N and θ, such that with probability larger than $1 - \overline{\mathfrak{c}}_2|\mathcal{H}_N|e^{-\alpha}$,

$$\sup_{h \in \mathcal{H}_N}\left\{\|\widehat{s}_h - s\|_\delta^2 - (1+\theta)\left(\|s_h - s\|_\delta^2 + \frac{\overline{s}_h}{N}\right)\right\} \leqslant \frac{\mathfrak{c}_2(1+\alpha)^3}{\theta N}$$

and

$$\sup_{h \in \mathcal{H}_N}\left\{\|s_h - s\|_\delta^2 + \frac{\overline{s}_h}{N} - \frac{1}{1-\theta}\|\widehat{s}_h - s\|_\delta^2\right\} \leqslant \frac{\mathfrak{c}_2(1+\alpha)^3}{\theta(1-\theta)N}.$$

On the one hand, for any $h \in \mathcal{H}_N$,

$$\begin{aligned}
&\|\widehat{s}_h - s\|_\delta^2 - (1+\theta)\left(\|s_h - s\|_\delta^2 + \frac{\overline{s}_h}{N}\right)\\
&\quad = \|\widehat{s}_h - s_h\|_\delta^2 + 2\langle \widehat{s}_h - s_h, s_h - s\rangle_\delta + \|s_h - s\|_\delta^2 - (1+\theta)\left(\|s_h - s\|_\delta^2 + \frac{\overline{s}_h}{N}\right)\\
&\quad = \|\widehat{s}_h - s_h\|_\delta^2 - \frac{1+\theta}{N}\overline{s}_h + 2W_{h,h} - \theta\|s_h - s\|_\delta^2
\end{aligned}$$

and

$$\|\widehat{s}_h - s_h\|_\delta^2 = \frac{U_{h,h}}{N^2} + \frac{V_h}{N} \tag{5.4}$$

So, with probability larger than $1 - \overline{\mathfrak{c}}_2|\mathcal{H}_N|e^{-\alpha}$, by Lemmas 5.1 and 5.2,

$$\begin{aligned}
\sup_{h \in \mathcal{H}_N}\left\{\|\widehat{s}_h - s_h\|_\delta^2 - \frac{1+\theta}{N}\overline{s}_h\right\} &\leqslant \sup_{h \in \mathcal{H}_N}\left\{\frac{|U_{h,h}|}{N^2} - \frac{\theta}{2N}\overline{s}_h + \frac{1}{N}|V_h - \overline{s}_h| - \frac{\theta}{2N}\overline{s}_h\right\}\\
&\leqslant \frac{2(\mathfrak{c}_{5.1} + \mathfrak{c}_{5.2})(1+\alpha)^3}{\theta N}
\end{aligned}$$

and then, by Lemma 5.3,

$$\sup_{h \in \mathcal{H}_N}\left\{\|\widehat{s}_h - s\|_\delta^2 - (1+\theta)\left(\|s_h - s\|_\delta^2 + \frac{\overline{s}_h}{N}\right)\right\} \leqslant \frac{\mathfrak{c}_2(1+\alpha)^3}{\theta N}. \tag{5.5}$$

On the other hand, for any $h \in \mathcal{H}_N$,

$$\begin{aligned}
&(1-\theta)\left(\|s_h - s\|_\delta^2 + \frac{\overline{s}_h}{N}\right) - \|\widehat{s}_h - s\|_\delta^2\\
&\quad = (1-\theta)\left(\|s_h - s\|_\delta^2 + \frac{\overline{s}_h}{N}\right) - (\|\widehat{s}_h - s_h\|_\delta^2 + 2W_{h,h} + \|s_h - s\|_\delta^2)
\end{aligned}$$

$$= -\theta\|s_h - s\|_\delta^2 + (1-\theta)\frac{\overline{s}_h}{N} - \|\widehat{s}_h - s_h\|_\delta^2 - 2W_{h,h}$$

$$\leqslant 2|W_{h,h}| - \theta\|s_h - s\|_\delta^2 + \underbrace{\left|\frac{\overline{s}_h}{N} - \|\widehat{s}_h - s_h\|_\delta^2\right|}_{=:\Lambda_h} - \frac{\theta}{N}\overline{s}_h$$

and

$$\Lambda_h = \left|\frac{U_{h,h}}{N^2} + \frac{V_h}{N} - \frac{\overline{s}_h}{N}\right| \quad \text{by Equality (5.4).}$$

By Lemmas 5.1 and 5.2, there exist two deterministic constants $\mathfrak{c}_3, \overline{\mathfrak{c}}_3 > 0$, not depending N and θ, such that with probability larger than $1 - \overline{\mathfrak{c}}_3|\mathcal{H}_N|e^{-\alpha}$,

$$\sup_{h\in\mathcal{H}_N}\left\{\Lambda_h - \frac{\theta}{N}\overline{s}_h\right\} \leqslant \frac{\mathfrak{c}_3(1+\alpha)^3}{\theta N}.$$

Moreover, by Lemma 5.3, with probability larger than $1 - 2|\mathcal{H}_N|e^{-\alpha}$,

$$\sup_{h\in\mathcal{H}_N}\{2|W_{h,h}| - \theta\|s_h - s\|_\delta^2\} = 2\sup_{h\in\mathcal{H}_N}\left\{|W_{h,h}| - \frac{\theta}{2}\|s_h - s\|_\delta^2\right\}$$
$$\leqslant \frac{4\mathfrak{c}_{5.3}(1+\alpha)^2}{\theta N}.$$

So, with probability larger than $1 - \overline{\mathfrak{c}}_2|\mathcal{H}_N|e^{-\alpha}$,

$$\sup_{h\in\mathcal{H}_N}\left\{\|s_h - s\|_\delta^2 + \frac{\overline{s}_h}{N} - \frac{1}{1-\theta}\|\widehat{s}_h - s\|_\delta^2\right\} \leqslant \frac{\mathfrak{c}_2(1+\alpha)^3}{\theta(1-\theta)N}. \tag{5.6}$$

Step 4. By Step 2, there exist two deterministic constants $\mathfrak{c}_4, \overline{\mathfrak{c}}_4 > 0$, not depending on N and θ, such that with probability larger than $1 - \overline{\mathfrak{c}}_4|\mathcal{H}_N|e^{-\alpha}$,

$$|\psi(h)| \leqslant \theta\left(\|s_h - s\|_\delta^2 + \frac{\overline{s}_h}{N}\right) + \left(\frac{\theta}{2} + \frac{2}{\theta}\right)\|s_{h_0} - s\|_\delta^2 + \frac{\mathfrak{c}_4(1+\alpha)^3}{\theta N}$$

for every $h \in \mathcal{H}_N$, and

$$|\psi(\widehat{h})| \leqslant \theta\left(\|s_{\widehat{h}} - s\|_\delta^2 + \frac{\overline{s}_{\widehat{h}}}{N}\right) + \left(\frac{\theta}{2} + \frac{2}{\theta}\right)\|s_{h_0} - s\|_\delta^2 + \frac{\mathfrak{c}_4(1+\alpha)^3}{\theta N}.$$

So, by Inequality (5.6) (see Step 3), there exist two deterministic constants $\mathfrak{c}_5, \overline{\mathfrak{c}}_5 > 0$, not depending on N and θ, such that with probability larger than $1 - \overline{\mathfrak{c}}_5|\mathcal{H}_N|e^{-\alpha}$,

$$|\psi(h)| \leqslant \frac{\theta}{1-\theta}\|\widehat{s}_h - s\|_\delta^2 + \left(\frac{\theta}{2}+\frac{2}{\theta}\right)\|s_{h_0} - s\|_\delta^2 + \mathfrak{c}_5\left(\frac{1}{\theta}+\frac{1}{1-\theta}\right)\frac{(1+\alpha)^3}{N}; \forall h \in \mathcal{H}_N,$$

and

$$|\psi(\widehat{h})| \leqslant \frac{\theta}{1-\theta}\|\widehat{s}_{\widehat{h}} - s\|_\delta^2 + \left(\frac{\theta}{2}+\frac{2}{\theta}\right)\|s_{h_0} - s\|_\delta^2 + \mathfrak{c}_5\left(\frac{1}{\theta}+\frac{1}{1-\theta}\right)\frac{(1+\alpha)^3}{N}.$$

By Inequality (5.3) (see Step 1), there exist two deterministic constants $\mathfrak{c}_6, \overline{\mathfrak{c}}_6 > 0$, not depending on N and θ, such that with probability larger than $1 - \overline{\mathfrak{c}}_6|\mathcal{H}_N|e^{-\alpha}$,

$$\left(1 - \frac{\theta}{1-\theta}\right)\|\widehat{s}_{\widehat{h}} - s\|_\delta^2 \leqslant \left(1+\frac{\theta}{1-\theta}\right)\|\widehat{s}_h - s\|_\delta^2 + \frac{\mathfrak{c}_6}{\theta}\left(\|s_{h_0} - s\|_\delta^2 + \frac{(1+\alpha)^3}{(1-\theta)N}\right); \forall h \in \mathcal{H}_N.$$

By taking $\theta \in (0, 1/2)$, the conclusion comes from Inequality (5.5) (see Step 3). □

Now, since

$$\widehat{f}_h = \widehat{s}_{h,(1,0)} \quad \text{and} \quad \widehat{bf}_h = \widehat{s}_{h,(0,1)}; \forall h \in (0, 1],$$

Theorem 5.3 allows to establish a risk bound on $\widehat{b}_{\widehat{h},\widehat{\mathrm{h}}}$, where $\widehat{h}$ (resp. $\widehat{\mathrm{h}}$) is defined by (5.2) with $\mathrm{v} = (0, 1)$ (resp. $\mathrm{v} = (1, 0)$).

Corollary 5.1 *Under Assumptions 5.1, 5.2 and 5.4, if $f(x), \delta(x) > m$ for every $x \in [\mathtt{l}, \mathtt{r}]$ ($m \in (0, 1]$, and $\mathtt{l}, \mathtt{r} \in \mathbb{R}$ satisfying $\mathtt{l} < \mathtt{r}$), then there exists a constant $\mathfrak{c}_{5.1} > 0$, not depending on N, $\mathtt{l}$ and $\mathtt{r}$, such that*

$$\mathbb{E}(\|\widehat{b}_{\widehat{h},\widehat{\mathrm{h}}} - b_0\|_{f,\mathtt{l},\mathtt{r}}^2) \leqslant \frac{\mathfrak{c}_{5.1}}{m^3}\Big[\min_{(h,\mathrm{h})\in\mathcal{H}_N^2}\{\mathbb{E}(\|\widehat{bf}_h - b_0 f\|^2) + \mathbb{E}(\|\widehat{f}_{\mathrm{h}} - f\|^2)\} + \|(bf)_{h_0} - b_0 f\|^2 + \|f_{h_0} - f\|^2 + \frac{\log(N)^3}{N}\Big].$$

Proof As in the proof of Proposition 5.4, and since $\delta(x) > m$ for every $x \in [\mathtt{l}, \mathtt{r}]$, there exists a constant $\mathfrak{c}_1 > 0$, not depending on N, $\mathtt{l}$ and $\mathtt{r}$, such that

$$\begin{aligned}\mathbb{E}(\|\widehat{b}_{\widehat{h},\widehat{\mathrm{h}}} - b_0\|_{f,\mathtt{l},\mathtt{r}}^2) &\leqslant \frac{\mathfrak{c}_1}{m^2}(\mathbb{E}(\|(\widehat{s}_{\widehat{h},(0,1)} - b_0 f)\mathbf{1}_{[\mathtt{l},\mathtt{r}]}\|^2) + 2\mathbb{E}(\|(\widehat{s}_{\widehat{\mathrm{h}},(1,0)} - f)\mathbf{1}_{[\mathtt{l},\mathtt{r}]}\|^2))\\ &\leqslant \frac{2\mathfrak{c}_1}{m^3}(\mathbb{E}(\|\widehat{s}_{\widehat{h},(0,1)} - b_0 f\|_\delta^2) + \mathbb{E}(\|\widehat{s}_{\widehat{\mathrm{h}},(1,0)} - f\|_\delta^2)).\end{aligned}$$

Moreover, by Theorem 5.3 and Proposition 5.5, there exists a constant $\mathfrak{c}_2 > 0$, not depending on N, $\mathtt{l}$ and $\mathtt{r}$, such that

$$\mathbb{E}(\|\widehat{s}_{\widehat{h},(0,1)} - b_0 f\|_\delta^2) \leqslant \mathfrak{c}_2(1 \vee \|\delta\|_\infty)\left(\min_{h\in\mathcal{H}_N} \mathbb{E}(\|\widehat{s}_{h,(0,1)} - b_0 f\|^2)\right.$$
$$\left. + \|(bf)_{h_0} - b_0 f\|^2 + \frac{\log(N)^3}{N}\right)$$

and

$$\mathbb{E}(\|\widehat{s}_{\widehat{\mathrm{h}},(1,0)} - f\|_\delta^2) \leqslant \mathfrak{c}_2(1 \vee \|\delta\|_\infty)\left(\min_{\mathrm{h}\in\mathcal{H}_N} \mathbb{E}(\|\widehat{s}_{\mathrm{h},(1,0)} - f\|^2)\right.$$
$$\left. + \|f_{h_0} - f\|^2 + \frac{\log(N)^3}{N}\right).$$

Therefore, there exists a constant $\mathfrak{c}_3 > 0$, not depending on N, $\mathtt{l}$ and $\mathtt{r}$, such that

$$\mathbb{E}(\|\widehat{b}_{\widehat{h},\widehat{\mathrm{h}}} - b_0\|_{f,\ \mathtt{l},\ \mathtt{r}}^2) \leqslant \tfrac{\mathfrak{c}_3}{m^3}[\min_{(h,\ \mathrm{h})\in\mathcal{H}_N^2}\{\mathbb{E}(\|\widehat{s}_{h,(0,1)} - b_0 f\|^2) + \mathbb{E}(\|\widehat{s}_{\mathrm{h},(1,0)} - f\|^2)\}$$
$$+ \|(bf)_{h_0} - b_0 f\|^2 + \|f_{h_0} - f\|^2 + \tfrac{\log(N)^3}{N}].$$

Let us conclude this section with some remarks about Corollary 5.1:

- By Corollary 5.1, the risk of $\widehat{b}_{\widehat{h},\widehat{\mathrm{h}}}$ is controlled by the sum of the minimal risks of

$$\widehat{s}_{h,(0,1)} = \widehat{bf}_h \quad \text{and} \quad \widehat{s}_{\mathrm{h},(1,0)} = \widehat{f}_{\mathrm{h}}\ ;\ (h, \mathrm{h}) \in \mathcal{H}_N^2,$$

  up to a multiplicative constant.
- A suitable choice for δ is the standard normal density function:

$$\delta(x) = \frac{1}{\sqrt{2\pi}} e^{-\frac{x^2}{2}}\ ;\ \forall x \in \mathbb{R}.$$

  First, as already mentioned, δ fulfills Assumption 5.4. Moreover, assume that $f(x) > m_1$ for every $x \in [\mathtt{l}, \mathtt{r}]$ ($m_1 \in (0, 1]$, and $\mathtt{l}, \mathtt{r} \in \mathbb{R}$ satisfying $\mathtt{l} < \mathtt{r}$). Since $\delta \in C^\infty(\mathbb{R}; (0, \infty))$, there exists $m_2 > 0$ such that $\delta(x) > m_2$ for every $x \in [\mathtt{l}, \mathtt{r}]$, and then

$$f(x), \delta(x) > m = m_1 \wedge m_2 > 0\ ;\ \forall x \in [\mathtt{l}, \mathtt{r}].$$

5.3.2.1 Control of U, V and W

This section deals with the control of the maps U, V and W introduced in the proof of Theorem 5.3 (see Lemmas 5.1, 5.2 and 5.3 respectively). The proof of Lemma 5.1 relies on the concentration inequality for U-statistics stated in Theorem 5.2, and the proofs of Lemmas 5.2 and 5.3 rely on Bernstein's inequality (see Theorem 5.1). Note that the properties of

$$\mathcal{K}_N = \{(\varphi, x) \mapsto \Phi_h(\varphi, x); h \in \mathcal{H}_N\}$$

provided in Proposition 5.6 are crucial throughout this section.

Lemma 5.1 *For every $h,\ \mathrm{h} \in \mathcal{H}_N$, consider*

$$U_{h,\ \mathrm{h}} = \sum_{i \neq k} g_{h,\ \mathrm{h}}(X^i, X^k)$$

where, for every $\varphi, \psi \in C^0([0, T])$,

$$g_{h,\ \mathrm{h}}(\varphi, \psi) = \langle \Phi_h(\varphi, \cdot) - s_h, \Phi_{\mathrm{h}}(\psi, \cdot) - s_{\mathrm{h}} \rangle_{\delta}.$$

Under Assumptions 5.1, 5.2 and 5.4, there exists a deterministic constant $\mathfrak{c}_{5.1} > 0$, not depending on N, such that for every $\theta \in (0, 1)$ and $\alpha > 0$, with probability larger than $1 - 5.4|\mathcal{H}_N|e^{-\alpha}$,

$$\sup_{h \in \mathcal{H}_N} \left\{ \frac{|U_{h,h_0}|}{N^2} - \frac{\theta}{N}\overline{s}_h \right\} \leqslant \frac{\mathfrak{c}_{5.1}(1+\alpha)^3}{\theta N}$$

$$\textit{and} \quad \sup_{h \in \mathcal{H}_N} \left\{ \frac{|U_{h,h}|}{N^2} - \frac{\theta}{N}\overline{s}_h \right\} \leqslant \frac{\mathfrak{c}_{5.1}(1+\alpha)^3}{\theta N}.$$

Proof First of all, note that for any $h, \mathrm{h} \in \mathcal{H}_N$,

$$U_{h,\mathrm{h}} = \sum_{i \neq k} (\overline{g}_{h,\ \mathrm{h}}(X^i, X^k) - \widetilde{g}_{h,\ \mathrm{h}}(X^i) - \widetilde{g}_{\mathrm{h},h}(X^k) + \mathbb{E}(\overline{g}_{h,\mathrm{h}}(X^i, X^k))) \tag{5.7}$$

where, for every $\varphi, \psi \in C^0([0, T])$,

$$\overline{g}_{h,\mathrm{h}}(\varphi, \psi) = \langle \Phi_h(\varphi, \cdot), \Phi_{\mathrm{h}}(\psi, \cdot) \rangle_{\delta}$$
$$\text{and} \quad \widetilde{g}_{h,\mathrm{h}}(\varphi) = \langle \Phi_h(\varphi, \cdot), s_{\mathrm{h}} \rangle_{\delta} = \mathbb{E}(\overline{g}_{h,\mathrm{h}}(\varphi, X)).$$

Since $\mathbb{E}(g_{h,\ \mathrm{h}}(\varphi, X)) = 0$ for every $\varphi \in C^0([0, T])$, by Theorem 5.2, there exists a universal constant $\mathfrak{m} \geqslant 1$, such that for any $\alpha > 0$, with probability larger than $1 - 5.4e^{-\alpha}$,

$$\frac{|U_{h,\mathrm{h}}|}{N^2} \leqslant \frac{\mathfrak{m}}{N^2}(\mathfrak{c}_{h,\mathrm{h}}\alpha^{\frac{1}{2}} + \mathfrak{d}_{h,\ \mathrm{h}}\alpha + \mathfrak{b}_{h,\ \mathrm{h}}\alpha^{\frac{3}{2}} + \mathfrak{a}_{h,\ \mathrm{h}}\alpha^2),$$

where the constants $\mathfrak{a}_{h,\mathrm{h}}$, $\mathfrak{b}_{h,\mathrm{h}}$, $\mathfrak{c}_{h,\mathrm{h}}$ and $\mathfrak{d}_{h,\mathrm{h}}$ are defined and controlled as follows:

- **The constant** $\mathfrak{a}_{h,\,\mathtt{h}}$. Consider

$$\mathfrak{a}_{h,\mathtt{h}} = \sup_{\varphi,\psi\in C^0} |g_{h,\mathtt{h}}(\varphi,\psi)|.$$

Since

$$\begin{aligned}\sup_{\varphi\in C^0} |\widetilde{g}_{h,\mathtt{h}}(\varphi)| &= \sup_{\varphi\in C^0} |\langle \Phi_h(\varphi,\cdot), \mathbb{E}(\Phi_{\mathtt{h}}(X,\cdot))\rangle_\delta| \\ &\leqslant \sup_{\varphi\in C^0} \mathbb{E}(|\langle \Phi_h(\varphi,\cdot), \Phi_{\mathtt{h}}(X,\cdot)\rangle_\delta|) \leqslant \sup_{\varphi,\psi\in C^0} |\langle \Phi_h(\varphi,\cdot), \Phi_{\mathtt{h}}(\psi,\cdot)\rangle_\delta|\end{aligned}$$

and, in the same way,

$$\sup_{\psi\in C^0} |\widetilde{g}_{\mathtt{h},h}(\psi)| \leqslant \sup_{\varphi,\psi\in C^0} |\langle \Phi_h(\varphi,\cdot), \Phi_{\mathtt{h}}(\psi,\cdot)\rangle_\delta|,$$

by Equality (5.7) and Cauchy-Schwarz's inequality,

$$\begin{aligned}\mathfrak{a}_{h,\mathtt{h}} &\leqslant 4 \sup_{\varphi,\psi\in C^0} |\langle \Phi_h(\varphi,\cdot), \Phi_{\mathtt{h}}(\psi,\cdot)\rangle_\delta| \\ &\leqslant 4\left(\sup_{\varphi\in C^0} \|\Phi_h(\varphi,\cdot)\|_\delta\right)\left(\sup_{\psi\in C^0} \|\Phi_{\mathtt{h}}(\psi,\cdot)\|_\delta\right).\end{aligned}$$

Then, by Proposition 5.6.(1),

$$\mathfrak{a}_{h,\,\mathtt{h}} \leqslant 4\left(\frac{\mathfrak{m}_\Phi}{h^3}\cdot\frac{\mathfrak{m}_\Phi}{\mathtt{h}^3}\right)^{\frac{1}{2}} \leqslant \frac{\mathfrak{c}_1}{h_0^3} \quad \text{with} \quad \mathfrak{c}_1 = 4\mathfrak{m}_\Phi.$$

So, since $h_0 \in [N^{-1/3}, 1]$,

$$\frac{\mathfrak{a}_{h,\,\mathtt{h}}\alpha^2}{N^2} \leqslant \frac{\mathfrak{c}_1\alpha^2}{N^2 h_0^3} \leqslant \frac{\mathfrak{c}_1\alpha^2}{N}.$$

- **The constant** $\mathfrak{b}_{h,\,\mathtt{h}}$. Consider

$$\mathfrak{b}_{h,\,\mathtt{h}}^2 = N \sup_{\varphi\in C^0} \mathbb{E}(g_{h,\mathtt{h}}(\varphi,X)^2).$$

Since

$$\begin{aligned}\sup_{\varphi\in C^0} \widetilde{g}_{h,\mathtt{h}}(\varphi)^2 &= \sup_{\varphi\in C^0} \langle \Phi_h(\varphi,\cdot), \mathbb{E}(\Phi_{\mathtt{h}}(X,\cdot))\rangle_\delta^2 \\ &\leqslant \sup_{\varphi\in C^0} \mathbb{E}(\langle \Phi_h(\varphi,\cdot), \Phi_{\mathtt{h}}(X,\cdot)\rangle_\delta^2)\end{aligned}$$

and

$$\mathbb{E}(\widetilde{g}_{\mathrm{h},h}(X)^2) = \mathbb{E}(\mathbb{E}(\langle \Phi_{\mathrm{h}}(X^1,\cdot), \Phi_h(X^2,\cdot)\rangle_\delta | X^1)^2)$$
$$\leqslant \mathbb{E}(\langle \Phi_{\mathrm{h}}(X^1,\cdot), \Phi_h(X^2,\cdot)\rangle_\delta^2) \leqslant \mathbb{E}\left(\sup_{\varphi\in C^0} \langle \Phi_h(\varphi,\cdot), \Phi_{\mathrm{h}}(X,\cdot)\rangle_\delta^2\right),$$

by Equality (5.7), Cauchy-Schwarz's inequality and Proposition 5.6.(1),

$$\mathfrak{b}_{h,\mathrm{h}}^2 \leqslant 16N\mathbb{E}\left(\sup_{\varphi\in C^0} \langle \Phi_h(\varphi,\cdot), \Phi_{\mathrm{h}}(X,\cdot)\rangle_\delta^2\right)$$
$$\leqslant 16N\mathbb{E}(\|\Phi_{\mathrm{h}}(X,\cdot)\|_\delta^2) \sup_{\varphi\in C^0} \|\Phi_h(\varphi,\cdot)\|_\delta^2 \leqslant \frac{\mathfrak{c}_2 N}{h^3}\overline{s}_{\mathrm{h}} \quad \text{with} \quad \mathfrak{c}_2 = 16\mathfrak{m}_\Phi.$$

So, for any $\theta \in (0, 1)$, since $h_0 \in [N^{-1/3}, 1]$,

$$\frac{\mathfrak{b}_{h,\mathrm{h}}\alpha^{3/2}}{N^2} \leqslant 2\left(\frac{\theta}{3\mathfrak{m}}\right)^{\frac{1}{2}}\left(\frac{\overline{s}_{\mathrm{h}}}{N^2h^3}\right)^{\frac{1}{2}} \times \left(\frac{3\mathfrak{m}}{\theta}\right)^{\frac{1}{2}}\left(\frac{\mathfrak{c}_2\alpha^3}{N}\right)^{\frac{1}{2}}$$
$$\leqslant \frac{\theta}{3\mathfrak{m}N}\overline{s}_{\mathrm{h}} + \frac{3\mathfrak{c}_2\mathfrak{m}\alpha^3}{\theta N}.$$

- **The constant** $\mathfrak{c}_{h,\mathrm{h}}$. Consider

$$\mathfrak{c}_{h,\mathrm{h}}^2 = N^2\mathbb{E}(g_{h,\mathrm{h}}(X^1, X^2)^2).$$

As already established,

$$\mathbb{E}(\widetilde{g}_{h,\mathrm{h}}(X)^2) \vee \mathbb{E}(\widetilde{g}_{\mathrm{h},h}(X)^2) \leqslant \mathbb{E}(\langle \Phi_h(X^1,\cdot), \Phi_{\mathrm{h}}(X^2,\cdot)\rangle_\delta^2). \tag{5.8}$$

Then, by Equality (5.7) and Proposition 5.6.(2),

$$\mathfrak{c}_{h,\mathrm{h}}^2 \leqslant 16N^2\mathbb{E}(\langle \Phi_h(X^1,\cdot), \Phi_{\mathrm{h}}(X^2,\cdot)\rangle_\delta^2)$$
$$\leqslant \mathfrak{c}_3 N^2\overline{s}_{\mathrm{h}} \quad \text{with} \quad \mathfrak{c}_3 = 16\mathfrak{m}_\Phi.$$

So, since $h_0 \in [N^{-1/3}, 1]$,

$$\frac{\mathfrak{c}_{h,\mathrm{h}}\alpha^{1/2}}{N^2} \leqslant \frac{\theta}{3\mathfrak{m}N}\overline{s}_{\mathrm{h}} + \frac{3\mathfrak{c}_3\mathfrak{m}\alpha}{\theta N}.$$

- **The constant** $\mathfrak{d}_{h,\mathrm{h}}$. Consider

$$\mathfrak{d}_{h,\mathrm{h}} = \sup_{(a,b)\in\mathcal{A}} \mathbb{E}\left(\sum_{i<k} a_i(X^i)b_k(X^k)g_{h,\mathrm{h}}(X^i, X^k)\right),$$

where

$$\mathcal{A} = \left\{(a,b) : \sum_{i=1}^{N-1} \mathbb{E}(a_i(X^i)^2) \leqslant 1 \text{ and } \sum_{k=2}^{N} \mathbb{E}(b_k(X^k)^2) \leqslant 1\right\}.$$

Since $X^1, \dots, X^N$ are independent copies of X,

$$\begin{aligned}\mathfrak{d}_{h,\mathrm{h}} &\leqslant \mathbb{E}(g_{h,\mathrm{h}}(X^1, X^2)^2)^{\frac{1}{2}} \sup_{(a,b)\in\mathcal{A}} \sum_{i<k} \mathbb{E}(a_i(X^i)^2)^{\frac{1}{2}} \mathbb{E}(b_k(X^k)^2)^{\frac{1}{2}} \\ &4N\mathbb{E}(\langle \Phi_h(X^1,\cdot), \Phi_{\mathrm{h}}(X^2,\cdot)\rangle_\delta^2)^{\frac{1}{2}} \\ &\quad\times \sup_{(a,b)\in\mathcal{A}} \underbrace{\left(\sum_{i=1}^{N-1} \mathbb{E}(a_i(X^i)^2)\right)^{\frac{1}{2}}}_{\leqslant 1} \underbrace{\left(\sum_{k=2}^{N} \mathbb{E}(b_k(X^k)^2)\right)^{\frac{1}{2}}}_{\leqslant 1}.\end{aligned}$$

Then, by Inequality (5.8),

$$\mathfrak{d}_{h,\mathrm{h}} \leqslant 4N\mathbb{E}(\langle \Phi_h(X^1,\cdot), \Phi_{\mathrm{h}}(X^2,\cdot)\rangle_\delta^2)^{\frac{1}{2}} \leqslant \mathfrak{c}_3^{\frac{1}{2}} N \overline{s}_{\mathrm{h}}^{\frac{1}{2}}.$$

So, since $h_0 \in [N^{-1/3}, 1]$,

$$\frac{\mathfrak{d}_{h,\mathrm{h}}\alpha}{N^2} \leqslant \frac{\theta}{3\mathfrak{m}N}\overline{s}_{\mathrm{h}} + \frac{3\mathfrak{c}_3\mathfrak{m}\alpha^2}{\theta N}.$$

Therefore, there exists a deterministic constant $\mathfrak{c}_4 > 0$, not depending on N, h and h, such that with probability larger than $1 - 5.4e^{-\alpha}$,

$$\frac{|U_{h,\mathrm{h}}|}{N^2} \leqslant \frac{\theta}{N}\overline{s}_{\mathrm{h}} + \frac{\mathfrak{c}_4(1+\alpha)^3}{\theta N}.$$

In conclusion, with probability larger than $1 - 5.4|\mathcal{H}_N|e^{-\alpha}$,

$$\sup_{h\in\mathcal{H}_N}\left\{\frac{|U_{h,h_0}|}{N^2} - \frac{\theta}{N}\overline{s}_h\right\} \leqslant \frac{\mathfrak{c}_4(1+\alpha)^3}{\theta N}$$
$$\text{and} \quad \sup_{h\in\mathcal{H}_N}\left\{\frac{|U_{h,h}|}{N^2} - \frac{\theta}{N}\overline{s}_h\right\} \leqslant \frac{\mathfrak{c}_4(1+\alpha)^3}{\theta N}.$$

□

Lemma 5.2 *For every $h \in \mathcal{H}_N$, consider*

$$V_h = \frac{1}{N}\sum_{i=1}^{N} g_h(X^i)$$

where, for every $\varphi \in C^0([0, T])$,

$$g_h(\varphi) = \|\Phi_h(\varphi, \cdot) - s_h\|_\delta^2.$$

Under Assumptions 5.1, 5.2 and 5.4, there exists a deterministic constant $\mathfrak{c}_{5.1} > 0$, not depending on N, such that for every $\theta \in (0, 1)$ and $\alpha > 0$, with probability larger than $1 - 2|\mathcal{H}_N|e^{-\alpha}$,

$$\sup_{h \in \mathcal{H}_N} \left\{ \frac{1}{N}|V_h - \overline{s}_h| - \frac{\theta}{N}\overline{s}_h \right\} \leqslant \frac{\mathfrak{c}_{5.2}(1 + \alpha)}{\theta N}.$$

Proof First of all, for any $h \in \mathcal{H}_N$, note that

$$\begin{aligned} \|s_h\|_\delta^2 &\leqslant \|\delta\|_\infty \|K_h * s\|^2 \\ &\leqslant \|\delta\|_\infty \|K_h\|^2 \|s\|_1^2 \leqslant \frac{\|\delta\|_\infty \|K\|^2 \|s\|_1^2}{h} \end{aligned} \tag{5.9}$$

and

$$\begin{aligned} \mathbb{E}(V_h) &= \mathbb{E}(\|\Phi_h(X, \cdot) - s_h\|_\delta^2) \\ &= \mathbb{E}(\|\Phi_h(X, \cdot)\|_\delta^2) + \|s_h\|_\delta^2 - 2\langle s_h, \mathbb{E}(\Phi_h(X, \cdot))\rangle_\delta \\ &= \mathbb{E}(\|\Phi_h(X, \cdot)\|_\delta^2) - \|s_h\|_\delta^2 = \overline{s}_h - \|s_h\|_\delta^2. \end{aligned} \tag{5.10}$$

Now, consider

$$\begin{aligned} v_h &= V_h - \mathbb{E}(V_h) \\ &= \frac{1}{N} \sum_{i=1}^{N} (g_h(X^i) - \mathbb{E}(g_h(X^i))). \end{aligned}$$

By Bernstein's inequality (see Theorem 5.1), for any $\alpha > 0$, with probability larger than $1 - 2e^{-\alpha}$,

$$|v_h| \leqslant \sqrt{\frac{2\mathfrak{v}_h \alpha}{N}} + \frac{\mathfrak{c}_h \alpha}{N},$$

where

$$\mathfrak{c}_h = \frac{\|g_h\|_\infty}{3} \quad \text{and} \quad \mathfrak{v}_h = \mathbb{E}(g_h(X)^2).$$

By Inequality (5.9) and Proposition 5.6.(1),

$$\mathfrak{c}_h \leqslant \frac{2}{3}\left(\sup_{\varphi\in C^0}\|\Phi_h(\varphi,\cdot)\|_\delta^2 + \|s_h\|_\delta^2\right) \leqslant \frac{\mathfrak{c}_1}{h^3}$$

$$\text{with}\quad \mathfrak{c}_1 = \frac{2}{3}(\mathfrak{m}_\Phi + \|\delta\|_\infty\|K\|^2\|s\|_1^2),$$

and by Equality (5.10),

$$\begin{aligned}\mathfrak{v}_h &\leqslant \|g_h\|_\infty\mathbb{E}(V_h)\\ &\leqslant \frac{3\mathfrak{c}_1}{h^3}(\overline{s}_h - \|s_h\|_\delta^2) \leqslant \frac{\mathfrak{c}_2}{h^3}\overline{s}_h \quad\text{with}\quad \mathfrak{c}_2 = 3\mathfrak{c}_1.\end{aligned}$$

So, for any $\theta \in (0,1)$, since $h_0 \in [N^{-1/3}, 1]$, with probability larger than $1 - 2e^{-\alpha}$,

$$\begin{aligned}|v_h| &\leqslant 2\sqrt{\frac{\mathfrak{c}_2\alpha}{Nh^3}\overline{s}_h} + \frac{\mathfrak{c}_1\alpha}{Nh^3}\\ &\leqslant \theta\overline{s}_h + \frac{(\mathfrak{c}_1+\mathfrak{c}_2)\alpha}{\theta Nh^3} \leqslant \theta\overline{s}_h + \frac{(\mathfrak{c}_1+\mathfrak{c}_2)\alpha}{\theta}.\end{aligned}$$

Therefore, with probability larger than $1 - 2|\mathcal{H}_N|e^{-\alpha}$,

$$\sup_{h\in\mathcal{H}_N}\left\{\frac{|v_h|}{N} - \frac{\theta}{N}\overline{s}_h\right\} \leqslant \frac{(\mathfrak{c}_1+\mathfrak{c}_2)\alpha}{\theta N}$$

and then, by Equality (5.10),

$$\begin{aligned}\sup_{h\in\mathcal{H}_N}\left\{\frac{1}{N}|V_h - \overline{s}_h| - \frac{\theta}{N}\overline{s}_h\right\} &= \sup_{h\in\mathcal{H}_N}\left\{\frac{1}{N}|v_h - \|s_h\|_\delta^2| - \frac{\theta}{N}\overline{s}_h\right\}\\ &\leqslant \sup_{h\in\mathcal{H}_N}\left\{\frac{|v_h|}{N} - \frac{\theta}{N}\overline{s}_h + \frac{1}{N}\|K_h * s\|_\delta^2\right\}\\ &\leqslant \frac{\mathfrak{c}_3(1+\alpha)}{\theta N} \quad\text{with}\quad \mathfrak{c}_3 = \mathfrak{c}_1 + \mathfrak{c}_2 + \|\delta\|_\infty\|K\|_1^2\|s\|^2.\end{aligned}$$

□

Lemma 5.3 *Consider*

$$W_{h,\,\mathfrak{h}} = \langle\widehat{s}_h - s_h, s_{\mathfrak{h}} - s\rangle_\delta;\ \forall h,\quad \mathfrak{h}\in\mathcal{H}_N.$$

Under Assumptions 5.1, 5.2 and 5.4, there exists a deterministic constant $\mathfrak{c}_{5.2} > 0$, *not depending on* N, *such that for every* $\theta \in (0, 1)$ *and* $\alpha > 0$, *with probability larger than* $1 - 2|\mathcal{H}_N|e^{-\alpha}$,

$$\sup_{h\in\mathcal{H}_N} \{|W_{h,h_0}| - \theta\|s_{h_0} - s\|_\delta^2\} \leqslant \frac{\mathfrak{c}_{5.3}(1+\alpha)^2}{\theta N},$$

$$\sup_{h\in\mathcal{H}_N} \{|W_{h_0,h}| - \theta\|s_h - s\|_\delta^2\} \leqslant \frac{\mathfrak{c}_{5.3}(1+\alpha)^2}{\theta N}$$

$$\text{and} \quad \sup_{h\in\mathcal{H}_N} \{|W_{h,h}| - \theta\|s_h - s\|_\delta^2\} \leqslant \frac{\mathfrak{c}_{5.3}(1+\alpha)^2}{\theta N}.$$

Proof For any $h, \mathrm{h} \in \mathcal{H}_N$,

$$W_{h,\mathrm{h}} = \frac{1}{N}\sum_{i=1}^{N}(\mathfrak{g}_{h,\mathrm{h}}(X^i) - \mathbb{E}(\mathfrak{g}_{h,\mathrm{h}}(X^i)))$$

where, for every $\varphi \in C^0([0, T])$,

$$\mathfrak{g}_{h,\mathrm{h}}(\varphi) = \langle \Phi_h(\varphi, \cdot), s_{\mathrm{h}} - s\rangle_\delta.$$

Then, by Bernstein's inequality (see Theorem 5.1), for any $\alpha > 0$, with probability larger than $1 - 2e^{-\alpha}$,

$$|W_{h,\mathrm{h}}| \leqslant \sqrt{\frac{2\mathfrak{v}_{h,\mathrm{h}}\alpha}{N}} + \frac{\mathfrak{c}_{h,\mathrm{h}}\alpha}{N},$$

where

$$\mathfrak{c}_{h,\mathrm{h}} = \frac{\|\mathfrak{g}_{h,\mathrm{h}}\|_\infty}{3} \quad \text{and} \quad \mathfrak{v}_{h,\mathrm{h}} = \mathbb{E}(\mathfrak{g}_{h,\mathrm{h}}(X)^2).$$

By Proposition 5.6.(1),

$$\begin{aligned}
\mathfrak{c}_{h,\mathrm{h}} &= \frac{1}{3}\sup_{\varphi\in C^0} |\langle \Phi_h(\varphi, \cdot), s_{\mathrm{h}} - s\rangle_\delta| \\
&\leqslant \frac{1}{3}\|s_{\mathrm{h}} - s\|_\delta \sup_{\varphi\in C^0} \|\Phi_h(\varphi, \cdot)\|_\delta \leqslant \frac{\mathfrak{c}_1}{h^{3/2}}\|s_{\mathrm{h}} - s\|_\delta \quad \text{with} \quad \mathfrak{c}_1 = \frac{1}{3}\mathfrak{m}_\Phi^{\frac{1}{2}},
\end{aligned}$$

and by Proposition 5.6.(3),

$$\mathfrak{v}_{h,\mathrm{h}} \leqslant \mathbb{E}(\langle \Phi_h(X, \cdot), s_{\mathrm{h}} - s\rangle_\delta^2) \leqslant \mathfrak{m}_\Phi\|s_{\mathrm{h}} - s\|_\delta^2.$$

So, for any $\theta \in (0,1)$, since $h_0 \in [N^{-1/3}, 1]$, with probability larger than $1 - 2e^{-\alpha}$,

$$
\begin{aligned}
|W_{h,\mathfrak{h}}| &\leqslant 2\sqrt{\frac{\mathfrak{m}_\Phi \alpha}{N}\|s_\mathfrak{h} - s\|_\delta^2} + \frac{\mathfrak{c}_1 \alpha}{N h^{3/2}}\|s_\mathfrak{h} - s\|_\delta \\
&\leqslant \theta \|s_\mathfrak{h} - s\|_\delta^2 + \frac{2}{\theta}\left(\frac{\mathfrak{m}_\Phi \alpha}{N} + \frac{\mathfrak{c}_1^2 \alpha^2}{N^2 h^3}\right) \\
&\leqslant \theta \|s_\mathfrak{h} - s\|_\delta^2 + \frac{\mathfrak{c}_2(1+\alpha)^2}{\theta N} \quad \text{with} \quad \mathfrak{c}_2 = 2(\mathfrak{m}_\Phi + \mathfrak{c}_1^2).
\end{aligned}
$$

Therefore, with probability larger than $1 - 2|\mathcal{H}_N|e^{-\alpha}$,

$$
\sup_{h \in \mathcal{H}_N} \{|W_{h,h_0}| - \theta\|s_{h_0} - s\|_\delta^2\} \leqslant \frac{\mathfrak{c}_2(1+\alpha)^2}{\theta N},
$$

$$
\sup_{h \in \mathcal{H}_N} \{|W_{h_0,h}| - \theta\|s_h - s\|_\delta^2\} \leqslant \frac{\mathfrak{c}_2(1+\alpha)^2}{\theta N}
$$

$$
\text{and} \quad \sup_{h \in \mathcal{H}_N} \{|W_{h,h}| - \theta\|s_h - s\|_\delta^2\} \leqslant \frac{\mathfrak{c}_2(1+\alpha)^2}{\theta N}.
$$

□

5.3.3 Basic Numerical Experiments

In this section, some numerical experiments on the adaptive 2-bandwidths Nadaraya-Watson of b_0 are presented for the three following models:

(A) $X_t = 0.5 - \int_0^t X_s ds + W_t,$

(B) $X_t = 0.5 + \int_0^t \sqrt{1 + X_s^2} ds + 2W_t,$

(C) and $X_t = 0.5 + \int_0^t \sqrt{1 + X_s^2} ds + \int_0^t (1 + \cos(X_s)^2) dW_s.$

For each model,

$$
\widehat{b}_{\widehat{h},\widehat{\mathfrak{h}}}(\cdot) = \frac{\widehat{bf}_{\widehat{h}}(\cdot)}{\widehat{f}_{\widehat{\mathfrak{h}}}(\cdot)} \mathbf{1}_{\widehat{f}_{\widehat{\mathfrak{h}}}(\cdot) > 0}
$$

is computed on $I = [-1.5, 1.5]$ from $N = 300$ paths of the process X observed along the dissection $\{\ell T/n; \ell = 0, \dots, n\}$ of $[0, T]$, where $n = 100$, $T = 5$, K is the standard normal density function, $\delta = 3^{-1}\mathbf{1}_I$, and both $\widehat{h}$ and $\widehat{\mathfrak{h}}$ are selected in $\{0.05\ell; \ell = 1, \dots, 6\}$ thanks to the PCO method (see (5.2) with $\mathrm{v} = (0, 1)$ and $\mathrm{v} = (1, 0)$ respectively). This experiment is repeated 100 times, and 5 adaptive PCO

Nadaraya-Watson estimations of b_0 are plotted on Fig. 5.1 for Model (A) and on Fig. 5.2 for Models (B) and (C). On average, as for the adaptive projection least squares estimator of b_0 (see Sect. 4.1.4), the MISE of $\widehat{b}_{\widehat{h},\widehat{\mathtt{h}}}$ is slightly increasing with the complexity of the model: 0.1567 (Model (A)) < 0.1786 (Model (B)) < 0.2666 (Model (C)). The same comment holds for its standard deviation: 0.0759 (Model (A)) < 0.1531 (Model (B)) < 0.2271 (Model (C)). This means that the more the model is complex, the more the quality of the estimation degrades, and it's visible on Figs. 5.1 and 5.2. However, for each model, the MISE of $\widehat{b}_{\widehat{h},\widehat{\mathtt{h}}}$ remains small.

Note that Marie and Rosier [7], Sect. 5.3 deals with some numerical experiments on the leave-one-out cross-validation (looCV) method for the 1-bandwidth Nadaraya-Watson estimator. As in the nonparametric regression framework (see Comte and Marie [4], Sect. 5), the looCV method seems numerically more effective than to select both h and $\mathtt{h}$ via the PCO method, but there is no theoretical guaranties on the associated adaptive Nadaraya-Watson estimator.

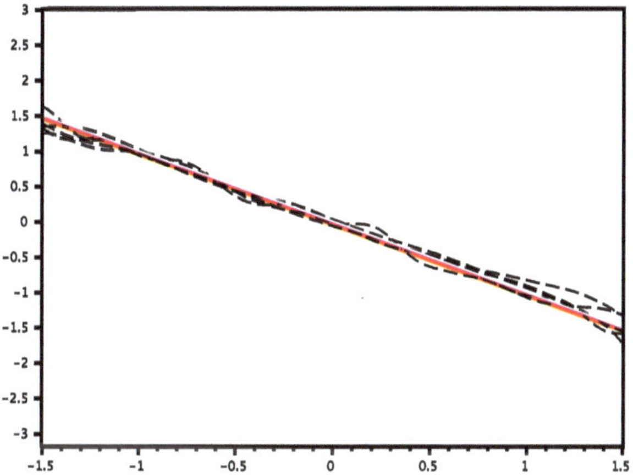

Fig. 5.1 Plots of 5 adaptive PCO Nadaraya-Watson estimations (black dashed lines) of b_0 (red line) for Model (A)

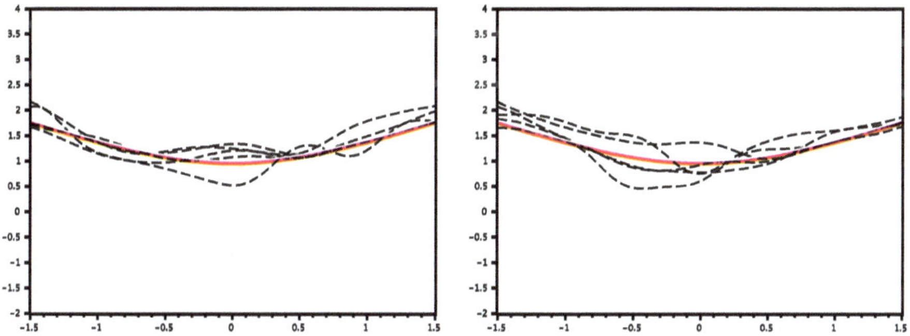

Fig. 5.2 Plots of 5 adaptive PCO Nadaraya-Watson estimations (black dashed lines) of b_0 (red line) for Model (B) (left) and Model (C) (right)

References

1. Kusuoka, S. and Stroock, D. (1985). Applications of the Malliavin Calculus, Part II. *Journal of the Faculty of Science*, University of Tokyo **32**, 1–76.
2. Menozzi, S., Pesce, A. and Zhang, X. (2021). Density and Gradient Estimates for Non Degenerate Brownian SDEs with Unbounded Measurable Drift. *Journal of Differential Equations* **272**, 339–369.
3. Comte, F. (2014). *Estimation non-paramétrique*. Spartacus IDH, France.
4. Comte, F. and Marie, N. (2021). On a Nadaraya-Watson Estimator with Two Bandwidths. *Electronic Journal of Statistics* **15**, 2566–2607.
5. Massart, P. (2007). *Concentration Inequalities and Model Selection*. Springer, Berlin-Heidelberg.
6. Giné, E. and Nickl, R. (2015). *Mathematical Foundations of Infinite-Dimensional Statistical Models*. Cambridge University Press, Cambridge.
7. Marie, N. and Rosier, A. (2023). Nadaraya-Watson Estimator for I.I.D. Paths of Diffusion Processes. *Scandinavian Journal of Statistics* **50**(2), 589–637.

Bibliography

1. Barron, A., Birgé, L. et Massart, P. (1999). Risk Bounds for Model Selection via Penalization. *Probability Theory and Related Fields* **113**, 301–413.

N. Marie, *From Nonparametric Regression to Statistical Inference for Non-Ergodic Diffusion Processes*, Frontiers in Probability and the Statistical Sciences,
https://doi.org/10.1007/978-3-031-95638-6

Index

N. Marie, *From Nonparametric Regression to Statistical Inference for Non-Ergodic Diffusion Processes*, Frontiers in Probability and the Statistical Sciences,
https://doi.org/10.1007/978-3-031-95638-6

FSC
www.fsc.org
MIX
Papier aus verantwortungsvollen Quellen
Paper from responsible sources
FSC® C105338

If you have any concerns about our products,
you can contact us on
ProductSafety@springernature.com

In case Publisher is established outside the EU,
the EU authorized representative is:
Springer Nature Customer Service Center GmbH
Europaplatz 3, 69115 Heidelberg, Germany

Printed by Libri Plureos GmbH
in Hamburg, Germany